The *Science of*
Consequences

The Science of
Consequences

HOW THEY
AFFECT GENES, CHANGE THE BRAIN,
AND IMPACT OUR WORLD

SUSAN M. SCHNEIDER

ILLUSTRATIONS BY RENÉ C. REYES

 Prometheus Books

59 John Glenn Drive
Amherst, New York 14228–2119

Medical disclaimer: Information related to medical conditions is intended only to be generally informative in nature. It cannot substitute for specific, personalized advice from qualified medical professionals.

Cover image © iStockphoto.com/Mazzzsur
Cover design by Jacqueline Nasso Cooke
Interior illustrations by René C. Reyes

Inquiries should be addressed to
Prometheus Books
59 John Glenn Drive
Amherst, New York 14228–2119
VOICE: 716–691–0133
FAX: 716–691–0137
WWW.PROMETHEUSBOOKS.COM
16 15 14   5 4 3 2

Library of Congress Cataloging-in-Publication Data

Schneider, Susan M., 1958–
    The science of consequences : how they affect genes, change the brain, and impact our world / by Susan M. Schneider.
        p. cm.
    Includes bibliographical references and index.
    ISBN 978–1–61614–662–7 (pbk. : alk. paper)
    ISBN 978–1–61614–663–4 (ebook)
    1. Reinforcement (Psychology) 2. Decision Making—Psychological aspects. 3. Nature and nurture. I. Title

BF319.5.R4S358 2012
153.8'5—dc23   2012023395

Printed in the United States of America on acid-free paper

For my mother
and in memory of my father

CONTENTS

PART 2: THERE'S A SCIENCE OF CONSEQUENCES?

PART 3: SHAPING DESTINIES

PREFACE

Every day our actions have consequences, large and small. A completed chore, a smile, a promotion. Consequences motivate: Newborns work to hear their mothers' voices. Toddlers graduate to turning lights on and off for that lovely, surprising feeling of control. A kaleidoscope of consequences awaits.

Despite their dazzling variety, consequences appear to follow a common set of scientific principles—which was a big surprise to early researchers. Correspondingly, very different consequences appear to share some similar effects in the brain. We've long known of so-called "pleasure centers," for example.

From simple rewards to far more complex relations, the science of consequences has expanded and flourished over the past century, becoming an integral part of psychology, biology, medicine, education, economics, and many other fields.[1] Taking an "interacting systems" approach, this book describes this science, its role in the larger realm of nature-and-nurture, and its many applications.

Consequences shape our choices, and our choices shape us and our societies. (Even the ancient Greeks understood that.) *The Science of Consequences* tells the story of how something so deceptively simple can help make sense of so much.

PART 1: CONSEQUENCES AND HOW NATURE–NURTURE *REALLY* WORKS

Imagine *not* being able to learn from consequences. A tiny primordial creature, for example, might have a reflex-like reaction moving it away from light. Move it to a lighted area to its food, then watch its automatic reaction kick in, shuttling it away and letting it starve to death in the midst of plenty. A little more flexibility would help a lot.

Once nature's ability to capitalize on success developed, it stuck. Birds and bees, even flatworms and fruit flies—we all learn from consequences. Which consequences? Some get learned the hard way, while others are taught or come naturally. That's a story in itself.

Part 1 explores the intricacies of the biological bigger picture, which is more complex than we could have imagined. We'll see nature and nurture always working *together*, interacting across all levels, from the nucleotide bases of our DNA to our rich, stimulating environments. Thanks to molecular biology, for example, we know consequences routinely activate and deactivate genes. Thanks to neuroscience, we can now see how learning from consequences expands and rewires the brain. From the latest neuroscience-based applications to the exciting new field of epigenetics, consequences are there.

PART 2: THERE'S A SCIENCE OF CONSEQUENCES?

Sometimes you get what you want quickly and easily: look out the window, see a view. More often, you have to work or wait: you might have to check several times for an eagerly anticipated e-mail, for example. As B. F. Skinner noted, consequences effectively come on some sort of schedule. Different schedules proved to have different orderly effects, effects powerful enough to influence how hard we work and how often. Thanks to schedules, *fewer* rewards routinely leverage *more* behavior. Changing a schedule can change the effect of a drug, even the very value of a consequence.

Schedules affect us every day (whether or not we realize it) and so do the signals that tell us what consequences are available. Likewise, when we decide what we "feel like" doing, we're choosing among consequences, immediate and delayed, positive and negative. Consider the way experienced procrastinators juggle conflicting consequences—something many of us know all too well. The science that applies to choice and decision making relies on consequences.

So do the language sciences. Ask for a smoothie at a juice bar and get one, voilà. Communication brings many rewards, and from baby babbling to birdsong, consequences have a surprisingly big say. Language lets us create rules, and they in turn create and destroy consequences. ("Don't eat yellow snow.") And naturally, we follow rules—or break them—because of consequences. Consequences bespeak even our inmost thoughts, as we'll see. They're everywhere.

PART 3: SHAPING DESTINIES

Because consequences are everywhere, their applications are too, from the home to the hospital, from the classroom to the boardroom. Few of us give or receive enough praise, researchers find. Yet something so simple can strengthen a marriage, rescue a struggling employee, and boost a child's self-esteem (if it's *earned*, and there's a tale). And that is just the beginning: The science of consequences enriches the lives of pets and zoo animals, fights prejudice, frees addicts of their addictive behaviors, and helps lift depression with its devalued consequences ("nothing to live for"). Ultimately, knowing what drives us puts us in the driver's seat.

Let's hope that is the case for many of our biggest societal challenges, which confront short-term against long-term *consequences*. In a world of instant gratification that tempts us to ignore later costs, perhaps we can use what we know to make better choices. We will all have to live with the consequences.

PART 1

CONSEQUENCES AND HOW NATURE–NURTURE *REALLY* WORKS

Chapter 1

CONSEQUENCES EVERYWHERE

*"An elderly male chimpanzee [was] observed in the field by
Dr. A. Kortlandt of Amsterdam's Zoological Laboratorium. It was
the chimpanzee's custom to go to a certain place where he could see
the sun go down. He went every evening and he would stay until the
sun had set and the color was gone from the sky. Then he would turn
away and find his place to sleep."*
—Sally Carrighar, *Home to the Wilderness*, 1973

Consequences provide the motivation that sends butterflies to flowers and people to the moon. The pursuit of happiness means the pursuit of consequences, large and small, sunsets included.

And consequences are everywhere. Some are immediate; others loom on the horizon to be anticipated or evaded. They're good, they're awful, they're everything in between. They work for tigers and for turtles—and for us. How ironic, then, that consequences and the science that focuses on them are so often overlooked.

Every day we work toward goals, goods, and incentives—consequences—for this minute, tomorrow, next year. Many rewards are obvious and immediate; others are subtle and easily missed. Some take a lifetime to achieve. Most are mundane, like paychecks and movies and smiles on friends' faces. Few remain unchanged in value, instead routinely transforming over time. Their variety seems infinite, far beyond biological drives like food, shelter, and sex. This chapter introduces that variety.

ORIGINS AND DEFINITIONS

Like most things, consequences started simple. We can't know what animal first learned from consequences, but flatworms, like the tiny planaria found in ponds, are reasonable candidates. Certainly both invertebrates and vertebrates are capable of this type of learning, and ancient planaria are considered common ancestors of both lines. On the planet for over 500 million years, they're the most primitive biological group to have "higher" neural features like brains (miniature ones, to be sure). Accordingly, their abilities have been researched extensively.

Despite looking like baby matchsticks that have taken to swimming, planaria can learn to work for consequences. In one study, some planaria had to move past an "electric eye" to shut off an unpleasant, intense light—which was a powerful reward. Others in a matched group were given the same intervals of light on and light off, but the intervals were independent of what these planaria did. Only the planaria with the light-off reward *dependent on their behavior* dramatically increased their interceptions of the electric eye—thus showing that this was truly learning from consequences and not just an effect of the light and dark alternation itself. These flatworms can even learn to work only when a signal is present: no signal, no reward.[1]

This research illustrates what *reinforcers* are: By definition, reinforcers both depend on behaviors and sustain them. If a behavior gets going and keeps going because of a consequence, that consequence is a reinforcer. If a behavior declines because of a consequence, that consequence is a negative (a punisher).[2]

Things that seem like rewards sometimes aren't: what matters is what actually happens, not the intention. Your Uncle Pete used to think that chucking you under the chin was rewarding. Wrong. Similarly, classroom reprimands sometimes function as reinforcers because of the attention that goes with them (from classmates as well as from the teacher).[3] If a "reward" has no effect on a behavior, then it's not a reward: Again, it's what actually happens that matters. I will use *reward* and *reinforcer* interchangeably in this book.

Because planaria are very simple animals, the consequences that are effective for them are limited. More complex invertebrates like the octopus show more sophisticated learning, influenced by a greater variety of reinforcers.

However, the range of effective consequences has mainly been explored in vertebrates, and it's large indeed.

WALTZING PIGEONS AND ROLLER-COASTER FISH: CONSEQUENCES ACROSS SPECIES

We love our pets when they behave like us—witness the dogs that enjoy watching TV and the owners who leave Spot's favorite program on. It may be harder to imagine common city pigeons doing something along these lines, but as a child, author and animal lover Gerald Durrell had a hand-raised pigeon that loved music and would snuggle close to the speaker of an old-fashioned record player. What's more, the bird performed distinctive dances to marches and waltzes.[4]

Later, scientists found that pigeons not only distinguish between different types of music, they also categorize unfamiliar tunes in the same way that people do, down to different eras of classical music.[5] And a number of species will work for music as a reward. The Java sparrow has been shown to prefer Bach over the twentieth-century dissonant composer Schoenberg, just like most of us. (Even rats feel the same way.)[6]

Ducks, such as common eiders, have been observed bobbing down rapids and then flying back up to repeat the fun. Otters slide down mudbanks, and starlings down snowy rooftops. Naturalist Edwin Way Teale once observed six chimney swifts soaring in a natural wind upflow, diving back down, and repeating the ride again and again. They were not feeding, they were "sporting in the wind."[7] What a different perspective on wild animals these episodes give us.

Even cold-blooded creatures get in on the act. I once laughed out loud at the sight of two puffer-type fish in an aquarium at an airport, of all places. They were taking turns deliberately swimming toward a strong circulation current that whirled them about and pushed them down, rather like a roller coaster. Other slimmer fish in the tank didn't appear to enjoy this and avoided the area. Perhaps the boxlike shape of the puffers changed their hydrodynamics. Or perhaps they were just well-fed youngsters with time on their hands.

What do these unusual animal reinforcers have in common with human rewards? A lot. The human love for roller coasters is based on reinforcers that are probably similar to those for the birds and the fish. Likewise, when it comes to musical harmony versus dissonance, the basic principles of hearing are similar in most vertebrates, even if the audible frequencies vary. Lacking ears, planaria are missing out.

GETTING STIMULATED: SENSORY CONSEQUENCES

But all creatures have sensory systems of some sort, even planaria, and they bring lots of rewarding possibilities.

As advertisers are well aware, in our constant-stimulation industrialized world, sheer sensory input is one of the most widespread reinforcers. Marketing experts outdo one another to hold our attention with rewarding visuals, sounds, and actions. Back in the nineteenth century, the first kaleidoscopes set sales records because of their novel combination of intrinsically reinforcing colors and shapes, which were perpetually complex, variable, and unpredictable—reinforcing on many levels (not unlike a computer screensaver). Even three-month-old babies will turn their heads so they can see complex patterns but not simple ones.[8]

Humans aren't the only ones to appreciate these features: in laboratory settings, mice and baby chicks will work to view more intricate patterns.[9] Chicks that had hatched just the day before, for example, already found looking at a design of stripes and semicircles more rewarding than eyeing plain gray.[10]

The level of optimal stimulation varies. High levels are sometimes rewarding and emotionally energizing, other times just stressful. After a period of high stimulation, even thrill seekers may prefer simple, quiet surroundings. "Boredom" indicates a lack of reinforcers, not a lack of stimulation.

Any small reinforcers, such as the view outside a window, can be fountains in a desert of tedium. But even when we're doing something we enjoy—

watching the Olympics or planting petunias—change is eventually welcome. Aristotle noted that "we cannot be moulded into only one habit, because each desire can operate only for short periods and must give way in turn to others."[11]

THE SPICE OF LIFE: VARIETY AS A CONSEQUENCE

Variety is usually a reinforcer, as formal research confirms, and not just for humans. Pets and zoo animals get bored too. Monkeys will work for the opportunity simply to view other monkeys, or a toy train, or even the normal room activity of people, and these consequences remain effective for long periods.[12] So monkeys like watching us as much as we like watching them—a lot. Animal behavior is so appealing that zoos and public aquariums enjoy big crowds, and millions of people have pets.

Variability is an important part of what is rewarding, in animal behavior and in plenty of other places. Sports are inherently variable; so are movie plots (well, good ones). Yes, we reread favorite books, but seldom immediately after we've finished them—and few would choose to watch A Hard Day's Night all day and all night, even though it's a classic. Some reinforcers can bear repetition, as when a child listens to the same song for hours (driving everyone else nuts!). But children eventually learn to ration that song, choosing variety instead.

Teachers do well to bear this in mind. A lack of variability in voice inflection can be the single best predictor of poor teaching evaluations.[13] Relatedly, I once had an excellent student who got A's on all his quizzes but tired of seeing comments in the mode of "Great" and "Nice job." He made up a list of alternatives like "Publish it" and "Hey-ho!"

The desire for variety can be subtle. We may not realize why a particular writer strikes us as dull despite interesting content, but monotonous sentence length and rhythm might explain it. As famed writing instructor Gary Provost noted, "The ear must have variety or the mind will go out to lunch."[14] In music, too, most successful composers across genres find ways to vary their melodies, harmonies, and rhythms to maintain interest. If that's not enough, music players offer a "shuffle play" feature that presents the tracks in random order, different each time.

The principle applies even to ordinary household tasks. From bestselling author and cook James Peterson: "I rarely prepare dishes according to an exact recipe because I never like to cook the same thing twice—I need to invent as I go along, or I get bored."[15] After the same chow three days in a row, even rats prefer new foods, new flavors.

THE CREATIVE CONSEQUENCE

What could be more intriguing than the discovery that variability is not only reinforcing but it is *itself* a reinforceable characteristic of our behavior? Creativity is not a zero-sum game, such that we each have only a limited amount. Instead, it's a "nurturable" that blossoms with encouragement: if you feed it, it will grow. And it's not limited to people.

Back in 1969, then dolphin trainer Karen Pryor and her colleagues knew better than most how inventive and adaptable dolphins are. They decided to see just how far dolphin creativity would go by rewarding only behaviors that they had never seen two particular dolphins do before. The dolphins met the challenge, coming up with quirky movements and stunts that they never would have had occasion for in the wild.[16] If dolphins could do this, how about kids? During the same period, scientists rewarded a few toddlers for building-block constructions that differed from prior designs—and the youngsters' creativity took off. Two of the three children came up with four times as many new designs when they were praised for doing so, compared to a baseline period with no particular consequences.[17]

Experimental psychologist Allen Neuringer's decades-long research program established the effect conclusively. His ordinary pigeons pecked two projecting buttons called keys. A reinforceable unit was eight key pecks that could be any combination of lefts and rights. To get delicious grains, the pigeons were able to produce sequences of eight that were consistently different from their last *fifty* sequences. And they weren't accomplishing this feat by memorizing. Instead, they behaved like random number generators, which was an efficient solution.[18] Can people do the same? First we have to overcome misconceptions of what "random" means mathematically. Neuringer rewarded participants for producing what they thought were random sequences of digits, then ranked the sequences on a number of statistical criteria and found that his volunteers did rather poorly. He discovered, for example, that we tend to alternate digits more often than is the case in a true random series. But with practice and feedback, people, like pigeons, were reliably able to generate random-appearing sequences, a skill that used to be considered extremely difficult.[19]

We may not often need to act randomly—quite the reverse. But if we're regularly reinforced—or punished—for variable behaviors like exploring or showing curiosity, guess what can happen? Just as with the kids' building-block designs, our creativity can soar—or plummet. Many Fortune 500 companies are quite aware of this, Google's innovation policies being one well-known example.

TAKING ADVANTAGE OF VARIETY

Some consequences for exploring, both positive and negative, are present naturally: variety, for example. We've seen how even little changes of scenery can be refreshing. Riding in a moving vehicle is reinforcing for those with good vision. Dogs get a kick out of the variety of scents to be sniffed on a short walk. On the other hand, familiarity is reinforcing under other circumstances, and exploring can mean leaving what is not only familiar but also safe. You don't find wild mice exploring in broad daylight but rather at night when they're at less risk; if their existing foraging locations fail them, they must seek new ones. They probably discovered their existing sources through exploring, so there's a history of reinforcement that helps.

The principle applies more broadly, of course. When do we challenge ourselves with the new instead of sticking to the old? Most of us like some sort of balance between the two, in the same way as optimal stimulation. Still, if we've had good luck (that is, been reinforced) seeing movies by particular directors, or reading books by particular authors, we're likely to stick with them. Unknowns, with their novelty and more questionable reward values, have a harder time vying for our attention.

Reinforcers like movies and books automatically offer variety. Even better, a "generalized reinforcer," like money, can be exchanged for a great variety of other reinforcers. Point systems take advantage of the power of these consequence choices. At one Michigan school, reward options included fun time with the principal for 75 points and free time in the computer lab for 100 points.[20] Researchers who study people often *have* to provide varied rewards to keep their participants motivated. A common solution is to offer lottery tickets for a chance to win prizes, such as gift certificates or pizzas. What about animals? Orcas (killer whales) at one SeaWorld never knew which reinforcers would come next: stroking and scratching, social attention, toys, and so on. The result was that "the shows can be run almost entirely without the standard fish reinforcers; the animals get their food at the end of the day."[21]

Animals with variety in their lives are healthier and happier, just like people. Environmental stimulants like toys are best varied, and those that produce unpredictable movements tend to be especially valuable and long-

lasting. One big success at one zoo was a sturdy swinging bag, hardly naturalistic, but rubbed and butted for hours by a rhino[22] (see chapter 13 for more
examples).

THE POSITIVE SIDE OF PROBLEMS

Monkeys can so love solving mechanical and other "educational" puzzles
that they will sometimes do so without any additional reward.[23] The activity
itself is intrinsically reinforcing. Karen Pryor reported on a challenge for a
rough-toothed porpoise: it had to learn to pick, among several choices, the one
that matched a sample. In the session where the porpoise first made notable
progress, it continued working even when it was no longer eating the fish it
had earned—as if it enjoyed something about the learning itself, just as we
do.[24] (Of course, social approval from the trainer certainly helped.) Along these
lines, here's an anecdote about biopsychologist Donald Hebb: A chimpanzee
working on challenging categorization problems hoarded its banana slices
until Hebb ran out. But the chimp continued to work—and after solving the
problems gave Hebb the banana slices![25]

Problem solving can be reinforcing for people too, whether it's locating
a leak, planning a party, or cracking a crossword. William Least Heat Moon
noted in his book *Blue Highways* that solving a maze is no fun if there are
no blind alleys.[26] On a broader scale, scientists are explorers who find and
solve different sorts of puzzles: "My reinforcers were the discovery of uniformi-
ties, the ordering of confusing data, the resolution of puzzlement," said B. F.
Skinner.[27] And from Nobel Prize–winner Linus Pauling: "Satisfaction of one's
curiosity is one of the greatest sources of happiness in life."[28] Thus, establishing a basis for curiosity helps promote happiness by creating new rewards.

TAKING CONTROL

Sometimes both variety and sensory stimulation are overshadowed. Rats press
a lever even when they get nothing except movement and the quiet click of

a microswitch, rather like infants repeatedly shaking a rattle for the same sound.[29] It's true that the more movable these objects are and the more sound they make, the more behavior. But the simple reward value of a change in stimulation can be greatly enhanced by having control.

From the cradle to the globe, we like having our own way. As any parent can testify, once toddlers discover that fascinating device, the light switch, everyone around knows it. Indeed, long before they're old enough to reach that high, infants love this game too. Skinner once turned a lamp on and off whenever his infant daughter moved her arm, and she was shortly conducting a symphony.[30] In more formal human research, the power of having power has been confirmed. Are we all control freaks, deep down?

If we are, animals are too. In one of my favorite examples, monkeys reportedly enjoy making touch-sensitive mimosa plants fold up. How do we know? They go out of their way to touch the mimosas and then stay and watch them. (So do kids. So do I, for that matter.[31]) And consider nocturnal deer mice in a lab, turning lights off if they can. "If the light is automatically turned off every half hour, however, the mice turn it back on. Even though the mice have an aversion to bright lighting, having control over the illumination is sufficiently rewarding to override it." Talk about "reverse psychology." Given switches on either side of their cage that turn the light on and off, the mice will run back and forth to do so many times a day.[32] (Remind you of the toddlers?)

With findings like these, it's not surprising that research shows animals, like people, usually prefer free choice to no choice. In one study, this was the case for pigeons even after a lot of extra reinforcers were provided when the birds picked the no-choice option (one key) over free choice (two keys)—although they eventually gave in and "followed the money." The birds returned to the free choice option when the opportunity for reward was once again the same for both options.[33]

So a lot of us love having choices and being in charge. No wonder, given these findings and our own experiences, that numerous surveys show job satisfaction to be correlated with the degree of control we enjoy in our work. This freedom is often cited as one of the reasons for the choice of an academic career. Among the benefits, a well-known study of British government officials found a strong relation between job control and health.[34] Having even minimal control is so powerful that nursing-home residents with plants in their charge—hardly a life-changer—showed significant health and psychological benefits compared to a similar group with plants cared for by staff.[35]

Animals benefit from having control too. A caged black bear that was psychologically disturbed, pacing back and forth in stereotypical fashion, did not improve when a feeder was introduced that unpredictably dispensed food. Picture yourself in a room with little to do. What would help? In the wild, bears spend much of their time foraging, and they control when and where. The zoo management tried hiding the bear's food in the enclosure, so that the bear was able to choose when to work to find it. Success!—and a much happier bear.[36] Even this minor level of control made a difference.

Teachers find one of their own reinforcers is "empowering" their students—literally, giving their students power over a consequence, from being able to solve mathematical word problems to understanding a foreign language. It's an immense pleasure for the teachers when the metaphorical light bulb turns on for their students, and that's often evident whether the students are human or animal. According to Karen Pryor, when empowered in this way, horses prance, dogs bark excitedly, and elephants "run around in circles chirping."[37] Some of the reinforcing effect is due to gaining control: "I can do this myself!" "I can solve the problem," "I can make the reward appear," "I'm in control." And the teacher's behavior is likewise reinforced by the accomplishment.

The "eyeblink game" illustrates the enjoyable process of giving an animal companion the pleasures of control. Try it with a pet bird: wait for a blink, then immediately close your own eyes, holding for longer than usual before reopening them. Reinforce each eyeblink this way, and shortly you should see a marked increase in the bird's eyeblink rate—often within five minutes. Visually oriented animals like birds love this game, to the point that my pet budgie, Goldie, would initiate it by closing her eyes for an unusually long period of time when I was nearby, and then repeat the long blink until I noticed and responded by blinking in return.[38] It took me five days instead of five minutes to get my pet rat, Clover, to the same level, but like most mammals, rats rely more on other senses. They also don't visibly blink very often (they have a hard-to-see nictitating membrane). In addition, for many mammals, including cats and dogs, direct eye contact can be a signal of aggressiveness. If you can look lovingly into your pet's eyes, though, you should be able to enjoy this game. I have even successfully played it with a few animals in the wild, and I suspect it would work with infants as well. The feeling of communication is powerful.

Having control is cool, but even so, it's not always preferred. It may seem counterintuitive, but studies show that having too many choices can be a downer. After all, what if we choose the wrong one? And think of all the extra effort required for a wise decision. One recent study showed that the more choices in an optional pension plan, the lower the employee participation.[39] Sometimes we would rather limit ourselves or choose that others should choose for us. We relinquish control.

But "power corrupts" because it's a powerful consequence that carries with it other powerful consequences (qualifying it as another "generalized reinforcer"). No wonder that, like the other consequences explored in this chapter, it is effective across so many species in the animal kingdom. How did consequences become a major force? In the next chapter, we look at the origins of that power.

Chapter 2

CONSEQUENCES AND EVOLUTION: THE CAUSE THAT WORKS BACKWARD

"As we all know, the cricket wasps get their crickets to the hole ready to be pulled in and then go down for the final inspection of the chamber. In the experiment in question, [entomologist Jean-Henri] Fabre pulled the cricket away a few inches when the owner was down the hole. The wasp came out, saw the cricket lying in the wrong place and put it back again on the threshold. This meant, of course, that there must be another final inspection before dragging it in. During this second final inspection Fabre once more pulled the cricket away a couple of inches. Wasp emerged, fetched it back and went down for a third "last" look. Fabre again pulled the cricket away. We have here the beginnings of perpetual motion if only both parties can keep it up. . . .

And then—which shows how careful one should be before generalising—he came upon wasps of the same species who acted differently; who learned in three goes to change their tactics. The same game was played with several of them, and they all fell for it. But only, in each case, three times. When Fabre, the persevering man, pulled the cricket away the fourth time they came up as usual and dragged it back but this time there was no last look; they took it straight down. They had learnt."

—John Crompton, *The Hunting Wasp*, 1955

Wesaw in chapter 1 that even primitive flatworms like planaria can learn from consequences. So can sea slugs, roaches, and tiny fruit flies.

Much of what invertebrates like these do falls under the category of "instincts,"[1] but consequences play a more important role in their lives than anyone used to realize. Still, it's quite a leap from planaria to the consequences found throughout human existence. How did we get here from there?

The genetics and neuroscience provide some helpful clues that we'll examine in the next chapters. Lacking time travel, what other clues are there to the development of learning from consequences? It's a shame that behavior doesn't fossilize. Or does it?

DANCE OF THE BALLOONS

One morning shortly after World War II had ended, entomologist Edward Kessel noticed white dots apparently floating in midair about fifteen feet up. He grabbed a ladder and a butterfly net and found tiny male flies with bulging eyes carrying round, silky "balloons" as big as the flies were. As they hovered, each fly clung to the "handle" of an even tinier dead prey insect attached to its balloon.[2] The show-offs belonged to one of the hundreds of species of dance flies. Why were they dancing? You get one guess.

Over the years, scientists discovered an apparent progression of court-ship rituals in the dance-fly family, illustrating the changes over time as new species branched off. Some males merely snare victims and then show them off in the communal dance. Females pick a male and load up on nutrition during mating. In other species, the males wrap their gifts in silk. What Kessel observed is a stage further along, in which the inedible wrapping had considerably outgrown the snack. Still later came production of the balloon alone. Finally, over time, some flies eventually stopped making the balloons and instead snagged floating leaf pieces or flower petals, dancing with them above the water to attract the other sex. These courtship rituals are as close to behavior fossils as we're going to get.

The dance moves in this case are unlearned "instincts." Learning from consequences doesn't fossilize in this way, but looking at how it works in creatures like these may still tell us something about how it developed. Meanwhile, keeping the focus on instincts is worthwhile, because they aren't as fixed and mechanical as we used to think. And consequences can affect them in a range of interesting ways.

FLEXIBLE INSTINCTS

The late Gilbert Gottlieb's research offers a famous example of how flexible instincts can be. Gottlieb, a pioneer in nature-*and*-nurture research, studied mallard ducklings. In "precocial" species like this, the young are born ready to roam, and mom leads the brood to safety away from the nest. "Imprinting" based on sound proved to be more important for these ducklings than visual imprinting. The mother's imprinting call ("Stay near me") is characteristic of all mallards, and her ducklings don't have to learn to prefer it over the imprinting calls of other kinds of birds. But how do they acquire this preference?

Gottlieb knew that in the nest, the in-the-egg duckling embryos start cheeping to each other about a day before they hatch. These cheeps don't sound like mom's imprinting call, so they had been assumed to be unrelated. Gottlieb decided to check. He found that when the embryos did not hear any of these prehatching embryo cheeps, those newly hatched birds then failed to have the normal preference for the mallard imprinting call. Gottlieb was able to show that both the rhythm and the pitch of the cheeping had to be heard before hatching: only then would the ducklings choose the correct maternal call.[3] The development of the typical my-own-species preference depended on *talking eggs*.

Do consequences enter into this drama? Yes. As researchers showed half a century ago, both natural and artificial imprinting objects can be effective consequences that chicken chicks worked to see. In one classic study, the more effective of two imprinting stimuli was also more effective as a reward. Of particular interest, the imprinted objects were effective rewards not just for following the object as in nature but for a variety of unrelated behaviors, like pecking a key.[4]

As researchers realized the possibilities, they tracked down more flexibility in this and other instincts. Gottlieb, for example, found that mallard ducklings that simply listened to hours of the domestic *chicken* imprinting call shortly before and after hatching came to prefer it over their own species' call, even though they'd still heard their own prehatching cheeping. The system of genes and environmental input that normally produces the imprinting-call preference simply couldn't handle this amount of unusual stimulation. Imagine those ducklings in the wild following a chicken instead of their mother.

For our purpose, a particularly powerful example comes from research with the most common North American quail species, the beloved (and declining) northern bobwhite, which calls out its name each spring. Like ducklings, the baby quail are precocial: they can toddle within minutes after hatching, the better to escape foxes and hawks. And like the ducklings, they're attracted to the imprinting call of their parents, which includes the father in the quail family group.

Intrigued by the power of consequences, one of my colleagues separated newly hatched bobwhite chicks from their pals individually, causing them to give their separation call ("I'm lost! Come find me!"). Every time an isolated chick called, my colleague played as an immediate consequence the very different imprinting call of the Japanese quail, a closely related species. After only five minutes (typically about thirty consequences), each isolated bobwhite chick was returned to its home. Other bobwhite chicks were isolated for five minutes as they listened to the same number of Japanese quail imprinting calls at the same times, but regardless of what they were doing—that is, not dependent upon their "I'm lost" calls. These bobwhite chicks, which heard the Japanese call *not* as a consequence, retained the normal strong preference for their own species' call when tested the next day. In contrast, the chicks that heard the Japanese call *only as a consequence* had become indifferent (on average), the normal preference for their own species' call eliminated in just five minutes.[5]

SONGBIRDS

There are plenty of other examples of interactions between instincts and consequences. For young male songbirds in a lab, the model song they learn from can become a reinforcer that they will work to hear.[6] During the development of their song, consequences can also be critical in other ways.

A well-researched species is the brown-headed cowbird of North America, which used to follow the bison and now, for lack of its partner, often hangs around cattle. In its original niche, how could the eggs be incubated and the chicks reared to adulthood in the normal way? The roaming herds of bison didn't stay in one place long enough. Cowbirds developed an unusual strategy known as brood parasitism, in which the females lay their eggs in the nests of other species and simply leave them there.

That means the young grow up hearing cowbirds as just one species among many other songbirds. They develop their normal song through a combination of instincts and other factors—which include consequences. When young males sing a song the gals like, these females give a wing stroke, a sort of cowbird applause. The lads take note and build their songs accordingly.[7] Consequences can also help shape the song of some finches, it's recently been discovered: when an unpleasant sound was played only when a particular part of their song began, the finches learned to avoid that consequence by modifying that part—and only that part—of their song.[8]

Nest building offers another example of instinct-consequence interactions. Male village weaver birds, for example, take months to learn to build a nest sufficiently well constructed to attract a female. Their pitiful first attempts can fall apart soon after they are created.[9]

The range of relations is also evident in commonplace food foraging. Animal behavior experts Eytan Avital and Eva Jablonka pointed out that spiders don't need to learn to spin their webs, although in other facets of their lives they are capable of learning from consequences. Hedgehog babies appear to learn based on their own experiences with what's easy to get and what's tasty—and as a result develop idiosyncratic foraging styles and food preferences. Partridge chicks peck where their mother points and are rewarded for doing so with edible tidbits. Surprisingly, at first many of them miss—pecking at a target also being learned through success and failure (consequences).[10] Once these chicks grab the goodie, though, they *don't* have to learn to like the taste.

BUGS THAT LEARN

If flatworms can do it, clearly the basic ability to learn from consequences doesn't require a large brain. Butterflies are giants compared to some of the planaria. Even so, it may be surprising to discover that butterflies too can learn from consequences. It's not an ability that comes to mind when we watch them gaily flitting from flower to flower in a sunlit meadow.

Is flower flitting affected by consequences? The pipevine swallowtail is a particularly attractive insect: the males boast glorious iridescent blue, while both sexes show off brightly patterned orange and blue underwings. In one study, Martha Weiss and her students let these butterflies loose on a field of yellow and magenta flowers. Sometimes only the yellow flowers had nectar, sometimes only the magenta ones (Weiss and company had fun arranging this). The butterflies quickly learned which color signaled the consequence they wanted.[11] While consequences were effective, these insects show an unlearned preference for the color yellow right from the start. As with the quail chicks, instincts, consequences, and learning coexist along a spectrum of relationships.

Later, Weiss observed wild wasps making short work of skipper butterfly caterpillars hiding in the leaf shelters they construct. Laboratory wasps viewing the shelters just seemed mystified. Could learning be involved? Indeed, research showed that only wasps that saw a caterpillar in its shelter and got to eat it learned to open the shelters for the tasty treats. Just seeing the shelters or the juicy bugs separately was insufficient.[12] The wasps had to learn that the leaf shelters signaled mealtime.

What about more arbitrary behaviors in small critters like insects? In the interests of science, many rats, mice, and even crabs have pressed many levers. It may sound whimsical, but a lone female stingless bee duplicated this feat. First, she learned to depress either of two tiny levers for nectar, and she did so reliably. Then a light was presented on either of the two sides, and only that lever would pay off. After six hours of training, the little insect pressed the correct lever 107 times, with only 33 errors.[13] Not bad—and a step forward in our understanding of learning from consequences. Later researchers built on these early studies to find very impressive learning abilities in honeybees (without using levers; see chapter 10).

When it comes to clever techniques, though, the method of Björn Brembs may take the cake. The tiny fruit fly *Drosophila melanogaster* is one of the best-studied organisms, the first insect to have its genome sequenced. Neuroscientist Brembs attaches the adult flies to a sophisticated flight simulator. By beating its wings, the fly doesn't move and yet it can "fly": through the magic of computers, the flies' turning torque produces an appropriate visual field that the flies can see. By adjusting their turning, the flies learned to avoid a beam of heat.[14]

WHICH CAME FIRST?

Invertebrates have both instincts and learning, so which developed earlier? At first glance, instincts may seem like the obvious choice, because they are less flexible and appear simpler to develop and operate. Is it even possible that the answer might have been learning?

When delay is risky, learning is a luxury: best to go with an unlearned method if you can, like imprinting. But imprinting and other "instincts" still rely on environmental inputs, as we've seen: Gottlieb's work with ducklings showed that hearing in-the-egg cheeping can be essential for the development of this "instinct." And my colleague's work with quail chicks showed how flexible such instincts can be in response to consequences: it took only five minutes to eliminate the chicks' normal imprinting call preference.

Start with those consequences instead and consider their flexibility. There is a natural fit between climbing a tree and eating its juicy fruit. Similarly, mother rats naturally trot over to retrieve their pups when they go astray. Would the chance to be reunited with a separated pup function as a reward for an arbitrary behavior like lever pressing?

First, pregnant mother rats in one study learned how to press a lever to receive Kellogg's Froot Loops. After the rats had given birth, they spent twenty-four hours with their pups. The young were then removed for twenty-four hours. Finally, the test: Would the moms work at pressing the lever to get their pups? Yes—if they actually got to pick up their babies, not just see them through a barrier. This reward method also worked for female rats given the hormones that would naturally be pumped into their systems after bearing young—illustrating the involvement of multiple components of the system.[15] (In another study, five dedicated moms each pressed a lever several *hundred* times within three hours, receiving a rat pup each time—not all hers!—and carrying it to her home.[16])

Outside the laboratory, many mammals and birds can learn arbitrary behaviors for the consequences they want. Reports from around the world show that graceful swallows, common house sparrows, and other birds have learned to trip electric eyes to open automatic store doors and forage or nest inside. Some fly into the path of the invisible beam, some sit nearby and lean over to trip it.

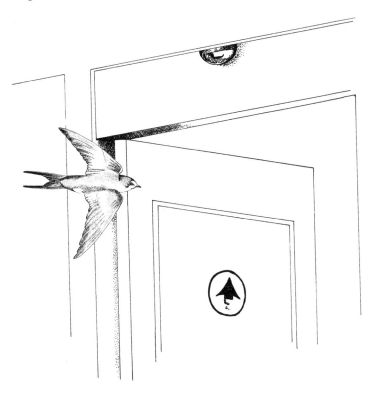

But those are vertebrates. Are invertebrates capable of producing arbitrary behaviors for rewards? The planaria that learned to intercept an electric eye provide an example. Stingless bees depress levers, as we've seen. In the fruit-fly flight simulator, avoiding heat worked even when the required flight pattern was arbitrary. Flying in circles to escape heat wouldn't work in the wild, but the little bugs had no trouble learning this maneuver.

The next step: As we saw, imprinted objects can reward arbitrary behaviors, and the natural imprinting process can be waylaid and overpowered by consequences. Imprinting is flexible. So it seems intuitively likely that instincts like this might branch out further over time and become full-fledged flexible learned behaviors. Quite so, but in the same spirit, Eva Jablonka and her colleagues have suggested that the opposite movement also occurs: learned behaviors can become instincts over time. It's simply a question of what is adaptive and hence gets favored over the course of natural selection.

How are these instinct/learning switches possible? In the mid-twentieth century, pioneering geneticist Conrad Waddington showed how the system can move back and forth, depending on what paid off. He raised fruit-fly maggots on a high-salt food, killing some. The young flies that survived were able to respond by producing bigger-than-normal structures on their rear ends to eliminate the salt. After breeding these adaptable flies together for several generations, he raised them on normal food for more generations. Some members of this line now showed the same structural change even without any exposure to salt: *assimilation* had taken place.[17]

It seems almost magical, but of course there's a logical explanation. Clearly, the capability to produce the salt-excreting structure had already been present in some individuals. A sufficiently salty environment simply triggered activation of the genes and other contributing factors required in these lucky individuals. Their descendants would then be expected to be more likely to produce the structure *without* the necessity of the salt stimulus—as indeed they were. It's not that big a difference, really, more a matter of what *signals* the creation of the enhanced rear end, since the capacity was already there (see chapter 3). As Waddington's experiments showed, selecting based on the former capacity (producing the structure only when the salt triggered it) means that the appearance of the latter also becomes more likely (producing the structure without the salt).

That's an anatomical difference, not a behavioral one—but the same principles apply. Gottlieb's research with mallard ducklings showed how hearing embryo cheeps before hatching was essential in developing the normal preference for the mallard imprinting call. Hearing these sounds changes the brain (see chapter 4). The cheeps are as much a part of the biological system as the genes are.

While we may never know for sure which came first, instinct or learning, we know a lot more about the interactions between these different behavioral processes. And that's helped scientists understand the course of evolution.

THE EVOLUTION OF CONSEQUENCES

Scientists have known for years that evolution is conservative, reconfiguring what's already available. Everything is potentially grist for the mill. As a result, many features evolved originally for purposes unrelated to their current function ("exaptations"). A classic example is the transformation of part of the reptile jaw into the bones of the mammal middle ear. Hey, the jaw bones happened to be in the right place and they could do the job, albeit not very well at first. But natural selection could take it from there.

A similar analogy for the development of learning from consequences may be provided by, of all things, feathers. At first glance, feathers may seem restricted to flight, but of course that's not the case: they have multiple functions. In fact, most bird feathers have nothing to do with flight. They offer temperature control and protection from the elements. The paleontology shows that insulating feathers first evolved from scales in land dinosaurs—not flying ones like pterosaurs but garden-variety dinosaurs, if such a thing can be said to have existed. Once the scales were present, they could be worked upon further by natural selection. Feathers on forelimbs grew and branched, allowing not only better climate protection but gliding and, ultimately, flight.

We can guess that in primeval times, only a few consequences may have been effective at first, and the ability to learn from them was very restricted. But with minor modifications and additions, consequences also took flight. Like feathers, consequences spread in greater variety, in turn motivating behaviors of greater variety. A critical step presumably occurred when consequences became effective for behaviors other than those that normally produced them. Lose your babies, rat moms? If they're near, go and carry them home. But if that doesn't work, learn to press a lever to retrieve them. Different behaviors, same function. As the list of effective behaviors and rewards lengthened, animals like dogs could enjoy catching a Frisbee instead of dinner. Generalized pleasure and pain centers developed in the nervous system and became busy places. Ultimately, as we saw in chapter 1, variability itself became a reinforceable characteristic of behavior.

Even planaria can learn about signals—ways to know when consequences are available. Add to that the ability to tolerate delays so consequences that aren't immediate can still be effective. Add generalization, the ability to respond similarly to events that are similar—something insects can do. Add the ability to form concrete categories like "berries"—not so different from generalization—and then abstract ones that still have something in common—"red"— and then arbitrary categories linked only by common consequences. (Some insects can handle abstract as well as concrete categories.) Add the ability to weight choices, so that bigger consequences count more and delayed consequences count less. Let consequences affect gestures and vocalizations. At this point, higher-level skills like communication and culture become not only possible but probable.

And it all started with those little flatworms.

BIRD BEAKS POINTING THE WAY: HOW CONSEQUENCES LEAD EVOLUTION

Not many birds can upstage fifty pink roseate spoonbills and a supporting cast of thousands of shorebirds, pelicans, gulls, terns, egrets, and herons— even a regal peregrine falcon cruising by. I had spent an hour paddling the mud flats of the Everglades, barely navigable at low tide even for my kayak. Now, bobbing in the water at a distance, hoping against hope, I was panning through all the spoonbills with my binoculars when I noticed a distinctly different pink bird in the background. I knew immediately that this was one of the holy grails of Florida birding: an adult greater flamingo. Paler, younger flamingos foraged nearby. I watched for about an hour, pinching myself. At closer range, I could even make out details of that unique boomerang-shaped bill.

The flamingo's bill is a classic example of behavior leading evolution. Irresistible little crustaceans of the briny bays reinforced the probing of the flamingo's predecessors, despite their originally clumsy beaks. Given enough time in this rewarding niche, genetically based structural changes in that beak followed.

Darwin's finches provide another famous example. In the Galapagos, the one pioneer finch species that settled these distant islands had all sorts of foraging opportunities, and "radiated" into over a dozen different species that often specialized in habitat, foraging style, or both. Foraging on big seeds, for example, created selection pressure for thick, husk-cracking beaks, because birds with bigger beaks had an advantage in survival and reproduction. For finches that pursued insects, thinner probing beaks were selected instead. "Sympatric speciation" —species separation without geographic isolation—is less common than geographically based speciation, but under the right circumstances, it can and does happen, and it clearly happened in the Galapagos.[18] If small-billed birds were more likely to mate with each other than with large-billed birds—perhaps simply due to spending more time in different habitats—then the stage was set.

The tables turned in these cases: once the ability to learn from consequences had evolved, it became an important *driver* of evolution. Different foraging styles for different food rewards in different habitats helped lead to different beaks. The behavior followed the consequences, and genetic change then followed the behavior—something even Darwin knew was possible. It's no different, in essence, from the way that the development of ancient oxygen-excreting bacteria brought our planet a whole new kind of atmosphere. Modern life could not have developed otherwise. Consequences also opened up whole new vistas as evolution proceeded.

The principle is pervasive and it continues to operate today. In the Pacific Northwest (as well as elsewhere), orca whales with different feeding habits have recently been suggested to be separate species. "Transient" orca pods roam widely after marine mammals like sea lions, while "resident" pods stay closer to shore and chase fish. "Offshore" orcas are like residents, but they travel much farther offshore. They all behave differently, and the three types—transient, resident, and offshore—avoid each other: Genetic analyses show that these groups of whales have not interbred for many generations, although they still could. Their dorsal fins look different, too.[19] Which change came first? Perhaps behavior again helped lead the way to ultimate genetic change.[20]

Some species go one step further. They don't just choose a niche because of the consequences found there, they *construct* the niche. Once in it, the same processes produce adaptation to that niche. Beavers are a classic example of busy niche constructors, but we can look a lot closer to home.

When people domesticated cattle, milk became a staple food—except that the ability to digest milk's lactose is usually lost after early childhood. What do you do if milk is all that's available in your herding society? Or you live in the northlands, layer clothing in the winter, and need the extra vitamin D that the sun no longer provides—but milk does? With enough variability in gene regulation—a change in timing is all that's required—and with enough selection pressure, it's not surprising that these clans developed extended lactose tolerance. But the behaviors and their consequences came first. They provided the critical selection pressure.[21]

THE CAUSE THAT WORKS BACKWARD

If behavior-led genetic change seems unexpected, it's only a consequence of the unusual way that both natural selection and behavioral consequences work—*selectionism*. It's not a straightforward process, like dominos. If you topple one domino, it topples the next, and on down the line: A causes B causes C, in order. The underlying principle of selectionism, however, is different: success is reproduced; failure isn't.

The force of selection is thus a cause that works backward, in a sense. Suppose you are eager to learn to make a free throw in basketball. Your initial efforts might be generally accurate but lacking in force, so your tosses fall short. Weak attempts disappear as a consequence, while forceful ones get rewarded with success, or at least balls that get close to the basket. (Hey, even the pros don't make every free throw.) The next time you shoot a few, you've gained some skills. Strong tosses are in, weak ones have been weeded out. The variations came first, the selection second, followed by the effect on your evolving basket-tossing repertoire.

Darwin noted that living things didn't just copy themselves once, they produced extra young that differed from each other. Finite resources meant that not all would be able to survive and reproduce, and "natural selection" ensured that those less fit (or just unlucky) lost the race. A blind robin hatches and then, as a consequence of its defect, perishes. As in learning from consequences, the variations come first, the selection second. Third is the effect on the robin gene pool: Genes that help to cause defects are weeded out. Genes that contribute to fitness—or are at least neutral—stay in.

Learning from consequences developed because it proved to help fit creatures for the everyday struggle of life: behaviors that succeed (that achieve reinforcement) are repeated, behaviors that fail disappear. Consequences modify behaviors across days and years, just as they alter species across centuries and millennia. Consequences in turn were selected. Injuries feel bad *because* animals that didn't avoid injury didn't survive. Pain, sad to say, was necessary for survival. But so was pleasure.

Take this one step further. Food tastes good *because* animals that liked eating survived. We ourselves don't have to learn to appreciate sugar or fats (more's the pity in these days of plenty for the developed nations). But our ancestors who found them rewarding were more likely to survive. The connection to the simple rewards of flatworms is evident.

B. F. Skinner was one of the first to note the parallels between the science he founded and natural selection in evolution.[22] Just as in evolution, there's both positive and negative selection. Positive selection increases the frequencies of advantageous genetic variants, as reinforcers strengthen the behaviors that produce them. Negative selection weeds out what causes harm, much as negative consequences diminish behavior. Neutrals are in between, sometimes flying under the radar for years.

It's easy to predict which dominos will topple. But how about predicting what will be a reinforcer or an adaptive feature? That's not quite as straightforward. Nonetheless, with sufficient knowledge, reasonably accurate predictions can be made. If your spouse loves vintage jazz (has a history of reinforcement), you can make a good choice for birthday show tickets. Similarly, if a moth lives on pine trunks, then new camouflage that blends in with the colors and textures of pines is likely to be successful—an effect that has been experimentally demonstrated in the wild as well as in the lab.[23] Research methods allow us to evaluate where the selection pressure really is and what's really reinforcing. That's where science comes in. (Otherwise it's altogether too easy to speculate, a mere parlor game.)

With such methods, researchers have been able to document the benefits of learning from consequences as well as moth camouflage. Bumblebee colonies that learned faster, for example, collected significantly more nectar than otherwise-similar colonies.[24] One short step for a bee, but a step toward the leaps that made our own skills possible.

Anthropologist Loren Eiseley called evolution "the immense journey." From butterflies to ducklings to dairy clans, consequences have shepherded these planetary wanderings, crafting bird beaks, new niches, and ever more complicated behaviors. Once living things could learn from consequences, life was never the same. A leading theorist recently argued that the evolution of learning from consequences may have caused the Cambrian explosion, the relatively sudden increase in the variety of animals famously discussed by Stephen Jay Gould in his book *Wonderful Life*.[25]

Wonderful consequences proved powerful enough not just to lead genetic change in evolution but to activate genes directly in our own lives.

Chapter 3

GENES AND CONSEQUENCES

"We share half our genes with the banana."
—Biologist Lord Robert May, past president of
Britain's Royal Society, 2000, quoted in *New Scientist*

How can we possibly share so much of our genetic material with the banana plant? What could we have in common? Cells. Metabolism. An orderly structure. Reproduction. Even sex (albeit rather different). More than most of us would have expected.

A protein called histone H4 forms part of our chromosomes, as it does in most species. The H4 sequence in pea plants is different from the mammal sequence by only two changes out of its 102 amino-acid building-block slots.[1] The gene that codes for H4 is obviously very similar across species. Our genome even has genes in common with single-celled creatures like yeast. According to Human Genome Project leader Eric Lander: "If there was a defect in how the yeast's cell division occurred, your gene would work just fine in the yeast, and people do that experiment all the time. We can do gene therapy to cure a sick yeast using a human gene."[2] Think of that the next time you break bread.

Compared to plants and yeast, our close kinship with a fellow mammal, the common house mouse, becomes understandable. "Greater than 99 percent of all genes in the human have a mouse counterpart, and vice versa. In fact, 96 percent of all genes in the human are found in the exact same relative order in human chromosomes as in the mouse chromosomes," noted geneticist Sean Carroll.[3] Learning from consequences is something we also have in common.

Because learning from consequences involves many genes as well as many other factors, we are only beginning to understand its particular genetic basis. A good place to start is by surveying what we know about genes, genomes, and the bigger system of which they are only a part.

MEET YOUR GENOME

We may share a lot of genes with simple species, but for complex creatures like us, surely nature went all out with the size of our genome, right? Humans were once thought to have as many as 100,000 genes, and as recently as the announcements of completion of the Human Genome Project, estimates were in the range of 30,000 to 40,000. We now know we have a mere 21,000 or so,[4] not much more than a nematode, a tiny primitive worm—and far fewer than some salamanders, grasshoppers, or even, believe it or not, single-celled amoebas.[5] Who would have guessed? It turns out that it's what is done with the basic materials that matters, not the number of genes. Indeed, as in many other species, most of our DNA (deoxyribonucleic acid) appears to have no function and is never expressed. According to current estimates, less than 2 percent of our human DNA actually codes for proteins. Another small portion is regulatory.

Like a frugal householder, nature has worked wonders with the materials and tools at hand, tinkering, in effect—just as with the exaptations illustrated by the mammalian ear bones of chapter 2. "Evo devos" (evolutionary developmental biologists) have pinned down which genes regulate the construction of the different body parts by turning protein-coding genes on and off—the master *homeobox* genes, for example. That makes these genes a prime source of tinkering. Like the gene that codes for the histone H4, the homeobox genes and other toolkit genes are very similar across most life forms. The *Pax-6/ eyeless* gene determines where eyes are placed not just in vertebrates (including us) but also in fruit flies and even flatworms. Considering how different insect and flatworm eyes are from ours, that's nothing short of amazing. The *distal-less* gene helps control the formation of appendages varying from the hundreds of "feet" in sea urchins to human legs, butterfly wings, and fish fins. When the genetic switches malfunction, mutant *Drosophila* fruit flies can have legs coming out of their heads.[6]

The proteins involved in body construction, like most proteins, are often used for other functions, too. As geneticist Michel Morange noted: "When the first genes affecting memory were isolated and characterized in *Drosophila*, there was surprise and disappointment when it turned out that the proteins coded by these genes were already well known for their role in metabolism. What makes a process specific is not the nature of its molecular components (and thus the genes that code for these components) but the way they are used and assembled in particular molecular pathways and specific structures."[7] This means that the genes that code for those proteins were regulated in different ways, in different locations, depending on these different uses. The maintenance of lactose tolerance selected for in the dairy clans of the last chapter is a good example. It's all a matter of tinkering.

GETTING TURNED ON

The toolkit genes are not the only way to turn coding genes on and off. All sorts of outside factors can serve as switches, including sight, sound, touch, diet, and behavior. This sort of genetic switching has been demonstrated in everything from nematodes (tiny roundworms) to canaries to people, for all sorts of genes.[8] For example, just looking at things changes genetic activity in the visual cortex of many mammals soon after birth.[9]

It's possible to go a step further and actually change genes themselves. If you take a bet (it had better be a big one) and stand near an unshielded pile of nuclear waste for a few seconds, the radioactivity might mutate your genes. You can also directly alter your genes through chemicals, or through infection by viruses that use the enzyme reverse transcriptase to insert their genomes into yours. (One of the big surprises is how often this has happened: over 100,000 viruses have plunked their genetic codes into what is now our human genome.[10])

But for most of us, outside factors like consequences affect our genes only through activating or deactivating them—and that's a great deal of power.

Genes code for proteins, the building blocks of our bodies, but they don't *make* those proteins unless they receive instructions. When we exercise, for example, we do it for the consequences: to feel better, to lose weight, to avoid nagging by our significant others, or just to get from point A to point B. Within two hours, gene expression changes in the affected muscles, regulating their metabolism.[11] We can, in effect, turn on these genes at will. Another factor involving consequences is stress. Start a family fight or blow off a big test, and the increased stress can shut down genetic activity in your immune system, resulting in increased susceptibility to colds and other infections.[12]

THE GENETICS OF CONSEQUENCES

How do we pin down the genes involved in learning from consequences? Myriads of interacting components have to be picked apart.

"Immediate early genes" can be expressed just minutes after appropriate stimulation (including outside experiences like working for consequences). An immediate early gene called *c-fos* is active in the brain. When neurons fire, *c-fos* gets expressed, so tracking it is a way to track neuron activity.

In an early study, some newly hatched chicken chicks learned to peck at chopped egg whites (understandably a favored food). The egg whites were scattered on a pebble floor, and at first the chicks wasted effort pecking the pebbles or pecking at but missing the egg whites. As mentioned in the last chapter, accuracy is learned through consequences—success and failure. An experimental group had one session with just pebbles, and a second session with pebbles plus food and a lot of learning. A second experimental group had two sessions with pebbles plus food. These chicks were rocket scientists at accurate pecking by the end of the second session, so not much new learning was going on. One control group was stuck with just pebbles, while a second control group simply stayed at home.

The focus was the forebrain, the chick equivalent of our cerebral cortex. What happened to gene activity there for *c-fos* and another, related immediate early gene, *c-jun*? The pebbles-plus-food chicks that were just starting to learn showed the biggest increase in the two forms of messenger RNA (ribonucleic acid) produced by the two genes, followed by a smaller increase for the rocket-scientist group. The two control groups showed essentially no gene activity.[13]

In a follow-up, another study compared *c-fos* gene activation in the motor cortex for three groups of female rats: an "acrobatic" group that learned to negotiate an obstacle course for a reward, another group that exercised as much but did not have to learn, and an inactive control group. The motor cortex gene turned on *only* with learning from consequences—not from the mere exercise.[14] These studies help illuminate both the role of genes and the relevant neuroscience (see chapter 4).

Learning from consequences involves memory, perception, and other processes. Looking at a different gene, *dCREB2*, Tim Tully and his colleagues shocked fruit flies in the presence of a signal odor. In a simple maze, the insects then chose between that odor and a new, unpaired odor—different consequences. The bugs learned quickly and with high accuracy, choosing the new odor. Depending on exactly how the training was performed, learning still occurred but was not retained when normal *dCREB2* activity was artificially diminished.[15] Progress. (Interestingly, mammalian *CREBs* are similar to *dCREB2*.) And, more recently, in rats pressing a lever for water, genes called *BDNF* and *Arc* were active in the prefrontal cortex while initial learning and associated brain changes took place. These genes were less active during simple maintenance of the behavior.[16] More progress.

Lurching down this long research road, just as physicists had fun naming quarks (for example, *charm*), biologists can turn humorous (the *sonic hedgehog* gene). A gene called *Klingon* may be directly involved with the changes that underlie learning from consequences, and plenty of other genes have been suggested. As a Star Trek fan, I'm thrilled. (*Captain Picard*, anyone?)

INTERACTIONS EVERYWHERE

One thing we do know is that the whole system is churning: genes, cellular processes, hormones and neurotransmitters, environmental factors of all sorts, the whole shebang. A major misunderstanding about nature "versus" nurture has been that it's an either/or proposition in which genetic and environmental contributions to a behavioral or physiological outcome can be separated. Instead, it's always "nature *and* nurture"—always genes and environment working together. These days, the geneticists themselves assure us of that.

With other factors held constant, for example, a difference in a single gene appears responsible for a difference in fruit-fly eye color.[17] But that gene can't be taken to code *for* eye color, which is the result of many genes and environmental factors working together. Indeed, two eyes of different colors—a common condition in cats and rare but regular in people—can be caused by either genetic or nongenetic (including environmental) abnormalities, such as infections or exposure to iron. In one case of Horner's syndrome, the difficult birth of a twelve-pound baby was considered to have caused the condition.[18]

Similarly, "genetic" diseases, even the few single-gene ones, aren't clear-cut either. If you get stuck with the problematic genotype, you get the disease, right? No, not every time, anyway. Instead, you might get lucky with other genetic or environmental players that compensate for the problem. People with the same sickle-cell anemia mutation, for example, can have very different levels of the disease. In men, sometimes the only apparent symptom is sterility. Even more surprising, you can still get one of these diseases if you *don't* have the problematic genotype (Huntington's disease, for example).[19]

In such a complex system, there are often multiple pathways to the same outcome.

It gets even more interesting. We each have two copies of a gene, one from each parent. For most people, one copy of the normal version 3 of the *APOE* gene plus one copy of version 4 equals trouble—higher risk of coronary artery disease (other things being equal). Substitute version 2 for version 4 and you're better off. However, for people with high cholesterol, it's having version 2 that loads on the highest risk.[20] The effects of a one-gene change like this can thus vary widely, depending on everything else that's going on in the system. Because that's the case, knockout gene studies can have unexpected outcomes. Surprisingly often, when a "critical" gene is knocked out, nothing at all happens.[21] The system had alternative routes to accomplish whatever was necessary.

The "it's a system" message was brought home years ago by the research team of Mark Cierpial and Richard McCarty. The "spontaneously hypertensive" genetic strain of rats produces animals that exhibit the characteristics of human hypertension (high blood pressure). Seems straightforward, doesn't it? But the hypertension develops only when the young rats are raised by mothers of the same strain. If raised by normal mothers, they don't show it—and neither do normal youngsters raised by spontaneously hypertensive mothers.[22] Both genes *and* environment are essential. It's a system.

Here's another example that entailed everything but the kitchen sink. Biopsychologist Stephen Suomi ran a series of studies focused on the neurotransmitter serotonin, which is involved in learning from consequences, emotion, and much more (including digestion in your gut). Neurotransmitters are the chemicals that convey brain signals across the synapses, the gaps between neurons; they're critical in brain function. In some cases, the short form of the serotonin transporter gene can be troublesome.

Suomi found that being raised by fellow youngsters isn't any better for rhesus monkeys than it would be for kids. But it turns out that these rear-yourself monkeys did especially poorly if they were stuck with the "short stick" of that serotonin gene: other things being equal, short-formers were more likely to be impulsive and aggressive and to drink more alcohol (indicating that it was a bigger reward). Is the short form always problematic? Lo and behold, when short-form and long-form youngsters were raised by foster mothers, the short-formers often did *better*: they drank less, were less impulsive, and ended up higher in the dominance hierarchy.[23] The process is neither simple nor straightforward. Just as for the *APOE* gene, the effects of a genotype can vary considerably depending on what else is going on. We're talking big, complex systems here, full of interacting factors.

WHAT'S INHERITED—AND WHAT ISN'T

One source of confusion about these gene-environment-system complexities is a scientific term called *heritability*. It does *not* mean what it appears to mean.

Consider genetically varying seeds raised in a controlled environment that is identical for all of them. Any differences in the height of these plants must then be caused by genetic differences, because everything else is identical. The heritability for height is 100 percent in this case, because all the variation in height must be due to variation in the genes. Now consider seeds for the very same plant, all clones with *identical genes*, raised in environments that are not identical. Any height differences now must be due to environmental and other nongenetic variation, so heritability is 0. For the same trait in the same species, heritability can be 0, or 100 percent, or anything in between. It depends on the circumstances. (Heritability figures apply only to particular populations in particular circumstances and *never* to individuals.)[24]

Furthermore, in all these groups, plants obviously need soil, water, and sunlight to grow. They don't grow very high without genes, either. Heritability *sounds* like it refers to the degree to which a trait like height is determined by genes. It doesn't.

Consider now the case of cloned plants raised in different environments. If they happen to end up at the same heights, can we conclude that genes control height? Clearly not. Again, plants obviously need soil, water, and sunlight to grow. *All* the nature-and-nurture factors are contributing to height, and they're interacting in complex ways. Any gardener could tell you that.

It gets better. Our number of fingers and toes turns out to have very low heritability: most of the variation is due to accidents, not to differences in genes, that's why. But the wearing of earrings in 1950s America had *high* heritability: only females normally wore earrings then, explaining the genetic connection. Not that genes and the proteins for which they code have anything to do with earrings directly, of course.[25]

The (in)famous twin studies have frequently been misinterpreted as well, although that's mostly history now. Suppose identical twins raised apart both love men named Bill, punk rock, and mountain biking. Well, genes code for proteins, not a fondness for The Clash. What, then, explains such shared features? David S. Moore covers the main explanations in his book *The Dependent Gene*. They include growing up in the same era and sharing race, gender, age, and (usually) similar social class, schooling, and appearance.[26] Under these circumstances, the statistical chance of shared likes and dislikes is high even among completely unrelated people.

Moreover, dissimilarities tend to get overlooked in the uproar over the power of coincidence. Indeed, as David Shenk uncovered in his book *The Genius in All of Us*, two planned TV documentaries on surprising similarities in raised-apart identical twins had to be cancelled because, on closer examination, the twins were full of surprising *differences*.[27] Some identical twins raised apart don't even look related. In one case, for example, one twin was short and skinny, the other substantially taller and chubby.

Cloned animals might be expected to be more "identical," but here too the effects of all the other factors make themselves known. Even tiny, genetically identical nematodes raised "identically" are never identical.[28] (And we're talking 1,000 cells here—that's the whole kit and kaboodle for this animal.) Clones of mice and cats have varied in many ways, in color patterns as well as in behavior.[29] Even when fed the same diet, only some cloned mice became overweight, for example.

Just like the twins and the clones, people from the same gene pool can go their different ways. The Pima Indians form a genetically related group that inadvertently conducted a natural experiment: Members in the United States adopted a new, less healthy lifestyle than their relatives across the border in Mexico. The levels of obesity and diabetes in the American group became far greater, despite their shared genes.[30]

EPIGENETICS: NEW KID
IN THE NEIGHBORHOOD

Speaking of obesity and diabetes, we're just beginning extensive research on what may be a surprisingly important part of the nature-nurture system—epigenetics (roughly, "associated with genetics"). Epigenetics affects both of these medical/behavioral problems.

All of your cells carry the same genome, so how do skin cells know how to specialize? And how do they divide to produce more skin cells, not other kinds of cells? They do so through the expression of some genes and not others—that is, through gene-activation patterns. We've already seen how environmental events and behavior can activate or deactivate genes. The new science tells us about cellular epigenetic mechanisms that influence which genes are more likely to be active. The number of epigenetic studies has exploded since the turn of the millennium, quite a turnaround for an obscure area that smacked of science fiction at first.

A paramecium is a single-celled critter that moves around by waving its rows of cilia, little eyelash-like engines of propulsion that cover its body. These cilia are lined up like a cat's fur, with a preferred direction. In 1965, two researchers cut out some of these cilia and reinserted them turned around in the wrong direction, and the insert "took." While that was surprising in itself, the researchers were amazed to see this nongenetic change inherited across multiple generations.[31] Whoa.

While that weird form of epigenetics doesn't apply to us, most others do. Examples include mechanisms involving histones, such as the histone H4 mentioned at the beginning of this chapter. Then there is DNA methylation, in which a methyl group (CH_3, that is) attaches to cytosine or adenine, two of the four nucleotide bases that constitute the genetic code. The code itself is not altered—but the more methyl groups that latch onto a gene, the less likely that gene is to be expressed. DNA methylation is hence one of the epigenetic ways of influencing which genes actually produce proteins.

In one recent study, DNA methylation and histone epigenetic patterns were examined in eighty pairs of identical twins. Young twins had nearly identical patterns. But the older the twins were, the more differences were found, especially given different medical histories or life experiences. Differences in these epigenetic patterns averaged four times greater in the older twins.[32] Your epigenetic patterns can and do change in your own lifetime.

The causes of change can include what you eat. Baby queen honeybees aren't any different genetically from worker bees; they just get fed a special "royal jelly" that modifies their epigenetic patterns.[33] As a result, they are substantially larger than the workers, and of course they behave differently. This is not just some weird bug phenomenon, either. Scientists gave food rich in B vitamins (and thus methyl groups) to pregnant females of a particular mouse strain. DNA methylation patterns changed *in the youngsters*, and those changed patterns caused differences in the risk of obesity and diabetes, and also cancer. The young mice even had different-colored fur.[34]

Could something like this happen to us? It appears so. A well-known example originates in the tragic World War II Dutch Hunger Winter, in which thousands of people died of starvation. Pregnant women who survived the ordeal had been malnourished for months. Their children were born smaller and at greater risk for obesity and diabetes—and epigenetic switches were shown to be involved.[35] There is some evidence that the third generation might also be affected. Over 100 cases of transgenerational epigenetic transmission across a variety of species are now well documented, and that includes mammals, not just bugs and one-celled critters.[36]

Some aspects of epigenetics remain controversial; others are well established. Epigenetic involvement in cancer belongs in the latter category, and a human epigenetic drug for leukemia (cancer of the blood) has been available for nearly a decade. But the flexibility of epigenetic switches is still being questioned, along with some of the links to behavior. It's a hot area of research that will be closely watched.

Assuming the results hold up, where do consequences fit in? To begin with, in the case of obesity and diabetes, consequences influence our dietary choices. And we will see later that people at higher epigenetic risk for obesity could use consequence-based methods to address the threat before they become overweight (see chapter 12).

Consequences can also affect epigenetics more directly because stress does, and consequences are involved in stress. Tania Roth, David Sweatt, and their colleagues tried stressing out foster-mother rats: the animals were plopped down in an unfamiliar location with insufficient bedding and immediately given unfamiliar babies. What would you do? The babies bore the brunt, getting dropped, stepped on, or "roughly handled" far more than rats raised by unstressed mothers, and being licked substantially less (even though the licking is normally reinforcing for the mothers[37]). The youngsters showed consistent changes in their DNA methylation and behavior patterns: for example, they were less likely to explore and had stronger stress responses.[38]

But all is not lost: Michael Meaney and colleagues found that these sorts of DNA methylation patterns—and behavior patterns—could be reversed when disadvantaged rat pups were given extra licking and grooming by adult females (regardless of genetic relationship, just as in Roth and Sweatt's research).[39] It's far too soon to be sure, but some evidence suggests that similar factors may operate in people. The implications would be huge.

Meaney and colleagues also found that enrichment later in life may be another way to compensate for the disadvantages of early stress.[40] That finding highlights a key point: what matters, ultimately, is flexibility. Because genes are just one part of a big, complex system, we've already seen modifiability in unexpected places—and there is a lot more to come.

In the next chapter, I will discuss the neurophysiological flexibility that plays with all this genetics/epigenetics/nature-*and*-nurture flexibility—and the cavalry-to-the-rescue role of consequences to take full advantage of it.

Chapter 4

NEUROSCIENCE AND CONSEQUENCES

"I can't put it into words. It's just—I use my brain. I just thought it. I said, 'Cursor go up to the top right.' And it did, and now I can control it all over the screen. It will give me a sense of independence."

—Quadriplegic Matthew Nagle,
first recipient of the BrainGate Neural Interface System implant,
which let him control a computer and a robotic arm with his thoughts

Science fiction, meet real life. Just as amazing, "thought control" was pioneered with rats and monkeys that had been fitted with similar implants.[1] This chapter heralds one "impossibility" after another. For all the genetics, epigenetics, and other factors in the nature-and-nurture system, it's the brain's flexibility that most directly makes possible our immense behavioral flexibility.

But that is a two-way street. If consequences can affect genes, it should be no surprise that they change the brain. Stimulating environments full of learning can actually expand the brain, growing more neurons and synapses—thereby supporting more and faster learning. This may sound like more science fiction, but it's as real and empowering as controlling a computer with your thoughts.

ENRICHMENT ON THE BRAIN

Each human brain has up to 100 billion neurons, and each neuron up to 10,000 connections with other neurons. That may be more connections than stars in the universe—absolutely mind-boggling.

How can we make the most of our potential? The extra stimulation and learning opportunities that come with "enrichment" are a logical choice. Enriched environments of all sorts have long been known to offer both neurophysiological and behavioral benefits. They've been heavily researched as a result.

This research field started rather touchingly. Like me and other rat researchers, early biopsychologist Donald Hebb had a pet rat, which naturally lived in an interesting environment for a rat. Hebb decided to see how it would do in a maze, and it beat the pants off the lab rats.[2] The rest is history.

Among this large body of research, one of my own contributions provides some whimsy: Quail chicks ran a maze—not your typical enrichment study by any means. At the end of the maze were several rewards, including extra heat, a speaker playing the standard species imprinting call, and a group of fellow chicks cheeping invitingly. An honors student and I created an interactive playground in some of the chicks' home bins, with toys like cardboard tubes to toddle through, balls to roll, and blocks to jump on or dodge about. Other chicks made do with just food and water and each other. The enriched chicks were far more active in the maze and about twice as likely to find their way to the finish line.[3]

Just being handled briefly early in life can have surprisingly significant effects for a number of species. Of course, other changes often follow upon it—a sort of cumulative outcome. Handled youngsters might be treated differently for the next few days by their parents and peers, which then leads to other differences. It's like learning to throw a ball accurately when you're unusually young. You may become a leader in ball games as a result, more popular than you otherwise would have been, and so on. For my chicks navigating the maze, learning from playing with toys could have long-lasting influence, too. Just imagine them out in their native wilderness, competing with their unenriched lab pals.

Most researchers have focused on our mammal relatives, revealing a long list of enrichment benefits: faster learning, better memory, less anxiety, and enhanced immune systems, and that's just for starters.[4] Recent evidence shows that enriched environments can even correct for some of the effects of brain damage, early deprivation, and lead exposure.[5] (Simply being able to learn a schedule of reinforcement—a minimal form of enrichment—can reverse some of the effects of lead exposure; see chapter 5 on schedules.) Many of these benefits appear to apply to humans too. Underlying all this are new and faster neural connections, increases in brain volume and weight, and enhancements in the supporting players of the nervous system, like *oligodendrocytes* (nerve-insulating support cells that are almost as long as their name), involved in conduction-boosting fatty myelination about the long axons that extend from the bodies of neurons like connecting wires. A whole slew of electrical and chemical activities and brain-cell changes orchestrate our highest functions, and enrichment and learning appear to influence most of them.

Just as with the genetics of learning from consequences, all these factors intertwine, and scientists are picking their way through these lush jungles bit by bit, trying to weave coherent stories. Again, as for the genetics, differentiating between enrichment changes based on learning instead of simple activity or sensation or the many other simultaneous goings-on has required a long-drawn-out endeavor. We've made immense progress, though.

Neuroscientist Akaysha Tang knew, for example, that a little bit of novelty in infancy could affect behavior later. For just three minutes per day over their first three weeks, baby rats were taken to a different environment—and their moms gave them extra licking when they returned (something we know from the last chapter can make a difference in itself). Babies in the control group stayed with their moms. Each group was then raised in the same way. Months later, both groups did equally well at learning to choose between two smells (for a Froot Loop as a reward). But after a delay of about a week, the minimally enriched group remembered what they'd learned much better than the control group did.[6]

Were there brain changes too? *Long-term potentiation* is a sort of greasing of the neuronal wires, so to speak—neurochemical changes that bring easier firing and faster connections, part of what makes learning and memory possible. Tang's study showed more long-term potentiation in the hippocampus (one of the brain's centers for learning and memory), but only for the enrichees. These changes lasted a whole year. Not bad for three minutes of novelty now and then.

NEURONS AND CONNECTIONS

Changes in long-term potentiation are just the beginning of the biological bases involved. Consider something seemingly simple, like walking—nerves are firing and muscles contracting, while balance and direction and pace are maintained through coordinated visual and ear-based vestibular action. Learning from consequences relies on levels upon levels of such coordinated effort.

New discoveries are steadily being made, even paradigm-shaking departures. Until recently, for example, most scientists thought new neurons could develop only during infancy. Then researchers found many new neurons being born in *adult* mice and rats, especially when the animals were in enriched environments. The dentate gyrus in the hippocampus was one of these happy sites, and, lo and behold, people also generate new neurons there, and at high rates too.[7] This is true even in the elderly. (Don't rejoice too much, though: many of our neurons perish each day.)

As the research momentum built, the number of species with documented neurogenesis in adulthood and old age grew. In monkeys, new neurons appeared not only in the hippocampus but in the neocortex, the outer layer of the brain that's associated with higher functions. And again we're talking more than a few. In rats—*adult* rats—the hippocampus creates thousands of new neurons each day. The more enrichment, the more neurons.[8] There's still uncertainty about the exact role of neurogenesis in learning, but at least now we know it exists.

When it comes to the formation of new neural synapses, there appears to be no doubt. Here's one example showing why: Remember the "acrobatic" rats of the previous chapter? The immediate early genes *c-fos* and *c-jun* got turned on only after learning from consequences—not after mere exercise. Well, in addition to this gene activation, substantially more synapses per neuron in the brain's motor cortex were created, but only after learning. And there are plenty of studies of this nature. In a follow-up by some of the same researchers, rats learned the skill of plucking rodent delicacies as they rotated by. Substantially more new synapses were created in this way than after easy lever pressing that required similar effort but little learning.[9] As enrichment pioneer William Greenough and a colleague put it, "The most general conclusion that can be made confidently is that the brain is an extremely plastic [flexible] organ, the structure of which is exquisitely sensitive to experience."[10]

Another major step forward takes us back to science fiction. Believe it or not, individual neurons can themselves "learn" from consequences, after a fashion. In 1993, Larry Stein's research team took neurons from a rat's hippocampus and maintained them "in glass." When tiny amounts of the brain chemical dopamine were provided right after a neuron fired, the neuron fired more often. Given a short delay between the firing and the dopamine dose, the response wasn't as strong. When no dopamine was given, the neuron slowed down considerably.[11] In this way, some of the main characteristics of behavior that's reinforced also occur in individual neurons.

Amazingly, scientists can even monitor what happens to individual neurons while an animal is living normally—and learning from consequences. When physical movements are involved, there's a lot of noise (extraneous interference) in the system, though. Neuroscientist Michael Platt and his colleagues minimized this interference by having their rhesus monkeys look to the right or left, with orange juice as the reward.

When looking in one direction paid off more than the other, the corresponding reward-related neurons were activated more. When rewards were delayed, the activity of the neurons modulated down. Neurons fired faster for consistently high payoffs and slower for low payoffs. The history of reward was preserved in this way, as it were.[12] As Platt noted, "Most of our choices are guided by the pay-offs associated with different actions in the past."[13]

Neurons slowed down considerably when a scheduled reward didn't come. But they were particularly active for "surprise" rewards—they gave a sort of extra lurch, sitting up and taking notice.[14] The unexpected may really be the sweetest. (We can relate to that.)

REWARDING CHEMICALS: DOPAMINE AND ITS COUSINS

The chemical signals in the brain are also pieces of the physiological puzzle. Neurotransmitters like dopamine convey brain signals across the synapses. They generally serve multiple functions, including activities completely unrelated to learning and memory. Learning and memory, in turn, involve a bewildering number of neurotransmitters in addition to dopamine: serotonin, endorphins, glutamate, norepinephrine, GABA, cAMP, acetylcholine, and peptides of various sorts (and that doesn't exhaust the list, either).[15] Dopamine has been the most heavily researched.

Several Parkinson's disease drugs mimic dopamine. In the mid-2000s, some Parkinson's patients made the news when they developed gambling addictions. In each case, they were on one of these dopamine-mimicking drugs, and there appeared to be a connection. From one of the neurology journal reports, here's a typical story: "This 41-year-old, married computer programmer reported never gambling in his life. . . . Within 1 month of reaching a [particular] dose he described being 'consumed' with the need to gamble on the Internet, losing $5000 within a few months." After he quit taking the drug, he lost his desire to gamble, describing the change as "like a light switch being turned off."[16] Similar stories covered the gamut across age, gender, and background. In each case, reducing the dose or entirely eliminating the drug ended the problem. In some of these Parkinson's cases, other rewarded behaviors like shopping, eating, drinking, and sex also ramped up beyond the norm.

We may all juggle a far more delicate balance than we realize.

Other drugs that incite dopamine activity can be powerful rewards. It's not surprising that these drugs include illegal substances like cocaine, but also on the list are prescription drugs like amphetamines and the amphetamine cousin methylphenidate (Ritalin and its kin, frequently prescribed for attention deficit hyperactivity disorder).[17] Even our friends the tiny planaria can be reinforced by a dose of methamphetamine[18] (and their neurons work in fundamentally the same way ours do).

How do researchers pin down what's going on? One method is to provide a chemical that's known to affect the neurotransmitter receivers, called *receptors*, in the synapses. Different receptors specialize in different neurotransmitters, and they can be stimulated in the same way through the wonders of chemistry—or shut down. Given dopamine-opposing drugs that slow down dopamine activity, animals gradually stop trying for rewards that previously had been powerful. If the animals are just starting to learn based on these rewards, they get nowhere.[19]

Because of its dopamine-reducing effect, the drug baclofen (generic name) was tested as a possible treatment for heroin addiction. Rats seem to like the effects of heroin and will work to get it. Under baclofen, however, heroin lost much of its power as a consequence, and the animals stopped working for it, or at least slowed down a lot.[20] Baclofen has been used to help alcoholics quit drinking, and research on its value for the treatment of other drug addictions is in progress.[21]

In such a complex system, though, dopamine's role remains uncertain. Some research has found that "liking" can remain even under low levels of dopamine. It may instead be an attention-influencing side of dopamine that's a basis of the brain's reward system.[22] Dopamine may turn out to be no more a simple "pleasure chemical" than are parts of the brain that at first appeared to be simple "pleasure centers."

PLEASURE CENTERS

A diagram of the parts of the brain that support just the dopamine reward system—the mesocorticolimbic system—would include more than a dozen named regions. Mammalian brain features with some evidence of involvement in learning from consequences (not necessarily involved with dopamine) are too many to list. A few of the most commonly researched are the nucleus accumbens, hippocampus, medial forebrain bundle, medial ventral prefrontal cortex, ventral tegmental area, and the nigrostriatal area. Believe me, I could add many more. As neuroscientists Ann Kelley and Kent Berridge summarized: "Reward-related behavior emerges from the dynamic activity of entire neural networks rather than from any single brain structure."[23]

That being so, do different consequences still share some common effects? Since their behavioral effects are similar (as we saw in chapter 1), that wouldn't be surprising. Take negatives, for example. (Please.) In one study using functional magnetic resonance imaging (fMRI), people were scanned while they were playing a computer ball game with what they thought were other people (actually just computer programming). When the "others" rudely excluded the real participant, one of the pain regions of the brain activated—in conjunction with reported hurt feelings. Very much like physical pain.[24] For positives, the body of evidence is much larger, and similar activity in the "pleasure centers" has been found for rewards as varied as music, money, and gazing at attractive faces.[25]

The discovery of "pleasure centers" is one of the oldest findings in the neuroscience of consequences. More than half a century ago, James Olds and

Peter Milner quite accidentally noticed that direct electrical stimulation of part of the brain could be rewarding: rats freely moving about returned to areas where they had received stimulation. When this stimulation was provided as a consequence for pressing a lever, responding took off.[26] The rest was history. Goldfish, iguanas, and dolphins work for brain stimulation; so do newborn animals, and so do people. Even invertebrates such as snails do.[27]

The effect can be frighteningly powerful: The original rats would stimulate themselves "day and night." They starved themselves, even with food nearby, rather than stop working for the brain stimulation. Given the chance to lever press either for heat or for stimulation, they would let their body temperatures fall to the point of death (averted only when the researcher stepped in).[28]

Sounds kind of like an addiction, yes?

But it's not a simple or straightforward phenomenon. Indeed, in a strange twist, stimulating a "pleasure center" can actually be punishing. In one experiment from the early days, researchers recorded the times that rats gave themselves electrical brain stimulation. Then the same brain stimulation was given for the same durations at the same times regardless of what the rats were doing (that is, *not* as a consequence). Believe it or not, now the rats worked to *stop* the same stimulation that they had previously worked to experience.[29]

The plot thickens. If the "pleasure center" stimulation-addicted rats are moved away from the lever—forced to take a break—then they can stop, and they avoid the very thing they were frantically working for a few moments before. "During preliminary observation, we simply removed self-stimulating rats from the lever by hand. They struggled for a few seconds and then lost interest. If released while struggling, they returned to the lever; if released when they had ceased to struggle, they did not return to the lever."[30] This is definitely not what normally happens when animals work for other rewards.

How about people? Researchers found that humans with stimulating electrodes in some of the "pleasure centers" don't necessarily report much pleasure, let alone the euphoria that might be expected.[31] Might the power of a consequence be separate from its power to give us pleasure? Evidently yes. It's considered the difference between "liking" and "wanting."[32]

That actually fits reasonably well with our everyday experience. We do lots of things that are motivating—there are consequences driving us—but that don't produce much feeling of pleasure. These consequences are still powerful. Similarly, perhaps the closest thing to a consensus about the mesocortico-limbic dopamine system is that it's essential for the motivating effectiveness of a consequence but not for the sensation of liking it.[33]

Whatever the complexities of these distinctions, it's clear that there are multiple "pleasure centers" in the brain, whether or not they always deserve that name (and alas, "pain centers" as well). The neurons from several different brain regions (including the neocortex, amygdala, and hippocampus) all converge in the neostriatum. Some researchers think this might be where many of the connections between behavior, consequences, and cues are made, with changes in value modulated by the strength of all the other signals and the weight of the past.[34] Here our history of rewards and defeats might be memorialized in our living, adapting personal computers, our brains with their billions of neurons—our dreams and goals in the flesh.

THE SKY'S THE LIMIT: NEUROPLASTICITY AND REAL-LIFE APPLICATIONS

In the service of those dreams and goals, one of the most exciting discoveries has been the immense modifiability of the brain, another "impossibility" come to life. As pioneer neuroscientist Michael Merzenich told author and psychiatrist Norman Doidge, "I had all of these reasons why I wanted to believe that the brain wasn't plastic in this way, and they were thrown over in a week."[35]

Take a classic series of studies by Merzenich and colleagues. Adult owl monkeys heard the same sounds, but a "touch-learning" group and a control group heard them just as background noise. For both of these groups there was *no* expansion in the auditory cortex, which handles sound. The touch learners were rewarded for learning a task involving touch. Only the relevant part of their brain's sensory cortex supporting *touch* expanded. In contrast, "sound-learning" monkeys, hearing the same sounds, were rewarded for learning the differences between them. The area in their auditory cortex that corresponded to those frequencies expanded.[36]

The effects were large, and the relationship between learning from consequences and expansion of brain coverage was so direct it was a thing of beauty.

Learning from consequences also changes the brain in some supposedly "impossible" ways. Harvard neuroscientist Alvaro Pascual-Leone and his colleagues divided human participants randomly into two groups, and members of one group wore blindfolds twenty-four hours a day for five days. Meanwhile, all participants were rewarded for learning Braille—an intense touch-based task. Using two different measures, Pascual-Leone showed that, deprived of anything to see, the *visual* cortex of the blindfolded learners gradually switched over to handling touch—in just five days. These changes were too fast, he thought, to be accounted for by new neurons or even primarily by new synapses. Instead, existing areas of the brain simply changed their function and started firing together with different partners.[37]

If the visual cortex can handle touch in normally sighted people under the right conditions, what about in blind people? Indeed, the visual cortex can be activated when a blind person is drawing.[38] Since it constitutes a rather large portion of our brains, the pressures for making use of the visual cortex in the absence of sight are presumably large. Work with animals has confirmed these effects: for example, the auditory cortex of ferrets can switch over to handle vision—a phenomenon also seen in some deaf people.[39]

Because of this kind of flexibility, the power of consequences can literally reshape the brain after profound loss. Learning from consequences is the driving force behind a successful new therapy for victims of stroke and related diseases. After a stroke, one side of the body can be paralyzed. Constraint-induced movement therapy motivates use of a damaged limb by binding or restricting the *good* limb. Then lots of positive reinforcement shapes use of the disabled limb, building on whatever limited movement was originally possible.

In one recent study, researchers worked with two groups of middle-aged and elderly stroke victims, each starting with similar severe arm disabilities (as measured by a standard rating scale). Each person could also move the afflicted arm slightly, something that is possible for most stroke victims. Given standard therapy, the control group failed to improve at all over several months. But after just two weeks of intensive training, patients getting the new therapy could move their afflicted arms far more freely, using them in real-world activities of daily living such as eating and dressing. These enhanced skills were still holding up well two years later.[40]

Researchers have confirmed that, while these changes are occurring, undamaged parts of the brain gradually take over for the damaged areas, and brain motor coverage that had shrunk expands again ("use it or lose it"). It even proved possible for brain cells to switch over and handle an arm or a leg on the *same* side of the body (previously considered another "impossibility").[41]

A major randomized controlled trial of the therapy with over 200 patients was successful, with benefits again documented to last for two years (at least).[42] More recently, another randomized controlled study investigated what happens when the therapy isn't started until about a year and a half after the stroke. Encouragingly, the benefits were comparable.[43] Other studies suggest that the therapy can bring benefits even many years after a stroke.

No wonder the American Stroke Association has described this consequence-based therapy as "revolutionary."

Let's conclude with the accomplishment that started off this chapter. Using arrays of tiny microwires designed to be fitted in the brain, a research team developed the BrainGate Neural Interface System that revolutionized Matthew Nagle's life. The idea is that a behavior like moving a joystick produces a consistent activity pattern in relevant parts of the brain. Careful selection and monitoring can allow for a sufficiently good assessment, enabling an interface that is able to convert thought into action.

It all started with rats implanted with the BrainGate technology that, to everyone's surprise and satisfaction, were able to replace lever pressing for food with thought control.[44] Upping the ante, two rhesus monkeys learned to move a joystick to control a cursor on a screen for juice as the reward. Given challenging tracking tasks, moving the cursor left, right, up, and down became second nature. Meanwhile, the team read the monkeys' BrainGate output and learned what to look for.

The big day came and the joystick was removed. The computer was there; the cursor was there. Now the monkeys could get juice only by "thinking" about moving the joystick. Both animals made the transition successfully, controlling the cursor and handling the tracking simply with their thoughts.

Their control ultimately became almost as fast as with physical movements. (They soon learned not to bother moving their arms.)[45]

Quadriplegic Matthew Nagle took only four days to learn to do the same thing. In the year that he was able to use BrainGate, Nagle could use his thoughts to check his e-mail on his own, change the channel on his TV, and even move a prosthetic device and lift objects from someone's hand.[46] Thinking often has consequences, but they are not always so easy to see.

Neuroplasticity isn't unlimited, not by any means. But there's a lot more of it than we used to think. And its potential can be unleashed through the power of consequences.

B. F. Skinner and others noted long ago that neuroscientists studying behavior would not get far without partnering with experts who understood fundamental behavior principles. The science of consequences will be enhanced by the new neuroscience discoveries, but the principles of consequences hold just as they always have. The relationship is like that between chemistry and physics: The discovery of quantum physics did not suddenly invalidate long-established principles of organic chemistry or chemical thermodynamics. Some disciplines within chemistry were enhanced more than others, and it will be interesting to see how psychology is affected. Certainly neuroscience can help guide research on behavior, just as behavior principles help guide neuroscience research. We're still only at the start of a beautiful friendship.

The chapters in part 1 illuminate how essential a systems approach is to understanding nature-and-nurture: genes, past history, behavior, environmental factors of all sorts, "pleasure centers," neurotransmitters, long-term potentiation, synaptogenesis, neurogenesis, epigenetics, and other biological factors—everything working together. Newly revealed are reserves of tremendous flexibility previously undreamt of. The wonderful implications of this flexibility reverberate throughout the rest of this book.

PART 2

THERE'S A SCIENCE OF CONSEQUENCES?

Chapter 5

CONSEQUENCES ON SCHEDULE: SIMPLE PRINCIPLES WITH SURPRISING OUTCOMES

"Laroche loved orchids, but I came to believe he loved the difficulty and fatality of getting them almost as much as the flowers themselves."

—Susan Orlean, *The Orchid Thief,* 1998

A s we'll see, Laroche's reaction might well be explained by the most pervasive feature of consequences—the principles of earning them. We experience these principles every day and yet they're almost unknown. They're hidden in plain sight.

Some earning principles are simple, some seem like they should be simple (but aren't), and others that seem complicated can have simple effects. To start addressing these conundrums, recall what consequences are. Things that are caused by what we do often shout—like the consequences for aiming a hammer wrong. Less obvious or delayed consequences whisper. But all consequences *depend on* a behavior. By definition, we have some control over real consequences, which is part of their power (recall the toddlers turning lights on and off in chapter 1). That's the simple foundation of the complicated ways they work.

FALSE CONSEQUENCES

But nothing about consequences is really simple. Even detecting a real consequence can be surprisingly difficult. Accidental "false consequences" can mislead us. An anecdote circulated about someone who had bumped against an electric pole right before a New York power blackout. That person's impression of being the cause of the hullabaloo was striking.

No harm was done in that case, but other coincidental false consequences are different. A successful test or poker game may create a "lucky shirt" that survives long past the normal toss-this-rag expiration date—and anxiety if the shirt gets lost. Professional athletes may have no more of these superstitions than the rest of us, but theirs are more visible, and they can balloon. In his book *Believing in Magic*, Stuart Vyse notes that baseball star Wade Boggs ate chicken every day and performed a five-hour pregame ritual.[1] Big-name musicians and actors, even chess players, have been known to succumb to superstition.

To analyze how false consequences work, psychologists took to their labs. In one study, a mechanical clown dispensed marbles on schedule, either every fifteen seconds or every thirty seconds, no matter what. Three- to six-year-old children were told that the marbles could be exchanged for toys later on. Most children soon developed superstitious behaviors like touching the clown's face to "make" it give marbles.[2] Research shows that adults share this tendency to try to get control where they can't. In a similar study, a woman trying for points ended up repeatedly jumping and touching the ceiling.[3] Don't laugh—coincidences have an almost eerie power, and the tendency to superstition is universal.

Animals included. One scientist was training a rat on a long series of complicated behaviors for a classroom demonstration. He reinforced one action at about the same time that the rat sneezed, a behavior these animals seem to be able to do at will. The rat then kept repeating the sneeze with that behavior.[4] The students may have thought it was part of the act; the rat clearly did. I expect that similar "superstitions" occasionally surprise us in our cats and dogs.

Research verifies that just one random "reward" can temporarily strengthen whatever behavior had been happening.[5] It's understandable, after all, to test whether it *was* a real consequence of what we did, one that we can bring on again. The tendency to keep trying then makes that act more likely to be occurring if another random reward comes, strengthening the superstition. It's insidious.

In people, language is a big influence, of course. Other young children working with the mechanical clown were told that when the clown's nose turned red, pressing it would help produce marbles. Children in this group followed this advice even though, again, marbles rolled along no matter what they did. The study continued in short sessions for over a month, and many children pressed the red nose throughout, some enthusiastically pounding away as often as once a second.[6] Most presses were not rewarded with a marble, of course, but they still continued, just like those superstitious sports rituals do even when the athletes have a bad day. (We'll see later what happens when rules and consequences disconnect.) This example illustrates a counterintuitive principle that in this case applies to true consequences, too: *fewer* consequences can support *more* behavior.

CONSEQUENCES ON SCHEDULE

Sometimes a consequence occurs for every behavior and not at all without the behavior. Every time you turn to look out the window, you'll see a view; if you don't turn, you won't. But most of the time, it's not that simple. You might have to check a dozen times before you receive an e-mail you've been waiting for. If it's an important one, you'll just keep checking. Similarly, most paychecks demand a lot of work. Consequences come on a schedule, not whenever we want.

Knowing something about how these schedules work comes in handy. Consider this: to discourage your dog Spot from begging annoyingly at the dinner table, is the "cold turkey," no-more-tidbits approach best, or is it kinder and just as effective to stop rewarding her begging gradually? The research shows, rather sadly, that "cold turkey" wins. Otherwise, occasional rewards for begging teach that it pays to keep trying . . . and trying . . . and trying.

We humans also tend to persist when we've learned that our efforts will eventually pay off. Sometimes that's beneficial, as when we keep job hunting or dating despite a series of disappointments. But it can be exceedingly harmful, too. Gambling is a prime example of the power of rare and unpredictable rewards, a combination that can be "addictive." Having learned that a payoff will come, compulsive gamblers can't stop playing the odds and gambling their lives away. Similarly, a trawler's lament: "Fishermen are always subject to what I call the lottery-day syndrome, the hope that with the next set of the net they'll haul up the jackpot."[7] (Even when more fishing boats and fewer fish have crashed fisheries around the world.)

These examples are all work based. The simplest work-based schedule is the "fixed ratio," in which the ratio of behaviors to consequences is fixed and constant. In the lab, for example, pigeons will peck 170 times for each small portion of food. That's not as demanding as it sounds, since they easily manage 60 pecks per minute without breaking a sweat. (Consider what they go through fighting over scraps on city streets.) Upping the ante to *360 pecks* for each reinforcer doesn't work: it's just too much effort to be worthwhile. But with a minor change, so that 360 pecks are necessary *on average*—what a difference. The birds work steadily away, sometimes having to peck substantially more than 360 times, but other times getting food for fewer than 100.[8] Humans behave just the same, and this variable arrangement gives far more bang for the buck than a fixed requirement. When the very next try might bring success, it's hard to stop.

These "variable ratio"/variable work-based schedules are more common than fixed schedules in everyday life. Remember Spot, the begging dog? Reading the newspaper is rewarded in this way too: not every article is interesting, but there's sure to be some good ones as you skim through.

WORK-BASED SCHEDULES AND THE POWER OF UNPREDICTABILITY

Not surprisingly, both of these work-based schedules occur in the working world. Piece-rate work is a good example of the fixed-ratio schedule in action. Psychologist Jim Mazur once worked in a factory where $10 was paid for each 100 hinges made using a machine. Workers typically made 100, then took a fairly long break, then worked steadily through another 100, the standard pattern on this schedule ("break and run").[9] The end-of-ratio pause makes intuitive sense: after you finish each ratio, you have to start all over again knowing that it will take a certain amount of time and effort to do the next. (The larger social consequences deserve consideration too: if enough employees worked at too fast a rate, the bosses have been known to raise the required ratio for everyone.) A variable-ratio schedule for piecework can be viewed as less fair, but in other work situations it's the method of choice. Salespeople who work on commission have a lot of motivation: the more people they contact, the greater the likelihood of a sale—maybe the very next customer. It's a variable-ratio schedule because sometimes the third contact brings a sale, sometimes the thirtieth. Those fixed-ratio pauses disappear with this improvement in motivation. In the laboratory and in real life, most of us—human and animal—choose variable schedules over fixed.[10]

This discovery has been capitalized upon in many ways. One disabled boy was rewarded for doing physical-therapy exercises. The youngster considered the variable-ratio schedule "an enjoyable game," and it won handily as a way of encouraging workouts.[11] (It even beat the "continuous reinforcement" schedule, where every exercise repetition got rewarded.) Other researchers found that variable schedules could be as effective as continuous ones in programs rewarding addicts for staying off drugs.[12]

Even when a variable schedule produces fewer rewards than a fixed one, the bias toward variable often persists, and that can be problematic. Two unsung pigeons trained originally on a richly rewarding variable-ratio schedule. When given a choice between variable and fixed schedules, they chose variable even when the variable schedule was made very lean and unrewarding. When they started losing weight, the researcher got concerned and put them on a fixed schedule alone. But when the bad variable schedule was reintroduced, the birds still made the same starvation choice, gamblers to the core.[13] Our own irrational choices can mirror those birds'. We'll see many times how history can be destiny.

Whether they are servants or masters, schedules are ubiquitous, especially work-based schedules: doing homework, mowing the lawn, paying bills . . . all those tiresome jobs where you have to do the work to get the consequence. (Or, as a happier example, making friends.)

If the work is too hard, though, like the pigeons that refused to peck 360 times per reward, we quit if we can. If forced to continue, we, like other animals, show signs of stress, such as uncharacteristically taking a break in the middle of a run. Doctoral students writing dissertations are notorious for bogging down in despair (technically, *ratio strain*), and few of us can't relate.

CONSEQUENCES ON TIME

Fortunately, not all consequences are work based. Suppose you're listening for your favorite song on the radio while heading out to restock the refrigerator. Just waiting increases the chances of success for the song—but no food will appear unless you work to get it. What's more, the faster you comb the aisles in the store, the sooner you can grab your baked beans and bananas. But nothing you do can hurry that song you're anticipating: it's time based, not work based. Other everyday time-based examples are checking for that important e-mail (variable), the appearance of a funny line in a TV show (variable), waiting for the washing machine to finish (fixed), and mealtimes (it depends).

For animals, it's the difference between ambush hunting and active pursuit. A bobcat crouching near prime rabbit grazing ground will eventually be rewarded with the chance to pounce (variable). In this time-based ("interval") schedule, just waiting will bring the opportunity for reinforcement. A short-tailed shrew on the lookout for a juicy bug, however, has to cast about repeatedly until it turns up a meal. The more active the shrew, the faster it finds its dinner (work based). One practically ran over my foot once in a West Virginia forest, so intent was it on its hunting.

Work-based or time-based, fixed or variable, one-time-only consequences qualify too. Plan to visit a distant country, for example, and you will find yourself gradually making more preparations as your fixed day of departure looms. If you're flying on standby and could leave at any time, though, you can't afford to delay. Similarly, working for a degree is clearly work based in the number of course credits that must be accumulated, but it may be more complicated. You probably have a time limit to satisfy the degree requirements, for example, just as finishing a lengthy project by a deadline may earn you a bonus at work.

A time limit understandably speeds up behavior. It's a schedule of consequences in itself, in fact, a *limited hold*. That term can fit literally: Holding an elevator door open for too long causes loud beeping. We learn to let go in time to avoid this embarrassing consequence.

In response to each schedule type, distinct patterns of behavior like this emerge.[14] Among the work-based schedules, a variable ratio with the same average reinforcer rate as a fixed ratio will produce more behavior faster (in part because there is typically no pause after each reinforcer). Time-based schedules produce distinctive patterns too. And this is just the beginning. Schedules and their characteristic patterns can ultimately require sophisticated mathematics to describe.[15] Dozens of types have been discovered or invented. As we have seen, many schedules are found directly in nature. Other schedules are "laboratory analogs" of the complications of real life, corresponding to standard practice in other sciences.

The characteristic patterns of behavior don't have to be high-speed a-go-go. In one intriguing schedule type, *low*-speed behavior is reinforced.[16] The reward is available only after a minimum amount of time has passed, like a time-based interval schedule—but you can't try for the reward during the interval. Failing to wait long enough means that the interval resets and you have to wait through another delay (a kind of inverted limited hold—you lose the reward if you *don't* wait long enough). Skilled performers on this schedule, then, become adept at waiting until the right time to try for the reward. This isn't just a made-up laboratory challenge. A good real-life example used to be waiting for a flooded carburetor to clear before restarting an engine. If you didn't wait long enough, you'd reflood the carburetor and have to start over. My dad was a master at the lost art of getting it right the first time. The advent of fuel injection means that I now use a new example, that of asking for a favor. Many of us are happy to grant favors, but not a lot of them in a row!—a decent interval is required in between. Some search engines also require a minimum delay between searches and reset if you don't wait long enough.

Most animals can handle this type of schedule, and some do just what young children do to make the time go by: they use delaying tactics. One researcher found that, while it was waiting, a pigeon bobbed its head in front of the correct key to peck. Young children in a delay study did well *only* if they developed similar active movements as time-passers.[17] There is a resemblance here to superstition: the delaying tactic isn't required, but it gets followed by a "false consequence" if it's long enough, so it can be likely to be repeated. Pacing impatiently around the room isn't necessary, but it pays off. Older children, like adults, learn to time the waiting intervals, so they don't need delaying tactics. If you need to wait ten minutes, you just check your watch.

As you'd expect, our individual histories influence our reactions, and that can be especially clear during a transition. People used to these sorts of slow-paced schedules may take a while to adjust to faster-paced ones.[18]

PROGRESS AND PERSEVERANCE

As human expertise researcher Anders Ericsson and colleagues have noted, one of the most important contributors to high-level skills is perseverance. They're excited about discoveries from "research about different 'schedules of reinforcement' . . . and their relation to sustaining motivation and effort over long periods of time."[19] Perseverance can be shaped using the power of schedules. As we've seen, variable schedules are a natural to build perseverance.

Furthermore, real life is dynamic and so are schedules. When a one-year-old is learning to walk, for example, we are thrilled with any kind of wobbly uprightness at first, then a single step, then two or three—a work-based ratio schedule, to be technical. The reinforced ratio doesn't stay the same for long: it lengthens until walking is taken entirely for granted. The same idea applies to reading, from nursery books to *War and Peace*. Actually, acquiring almost any skill would qualify. These "progressive ratio" schedules work well to build perseverance as long as the requirements don't progress too quickly—or the dreaded ratio strain that we have already seen may shut things down.

Perseverance in the face of such stress is justly celebrated, but everyone eventually reaches a breaking point. That's fortunate. Stalwarts who are used to lean variable schedules can persist *too* far, as when a would-be marathoner of my acquaintance practically had to be hospitalized after his first attempt—going overboard, rather like the gamblers on their variable schedules. One glory of being human is overcoming extreme barriers to the consequences we most desire, but it can be perilously akin to stupidity. Wisdom is knowing when the consequence is worth the effort, and we each hope it will come with age.

The ability to handle the different schedule types would seem to take some maturity, too. But we quickly adjust to most of them—because we have to. Newly hatched chicks will peck many times to earn a small portion of food if that's what's required. And they show a typical pattern on a fixed time-based schedule with few pecks at first, building up to a lot at the end of each identical interval when the food becomes available.[20] Babies in a number of species show these standard patterns—including humans.[21] After all, these patterns are usually an efficient way of getting the consequence.

MAKING THE MOST OF SCHEDULES

How can we be blind to the schedules affecting us every day? In the same way that we overlook everyday consequences themselves—all too easily. Understanding how schedules work, though, means we can cope with them more intelligently: enjoying the reinforcer in exchange for less effort, while bracing for any negative effects or avoiding them entirely. It gets easier to escape getting hooked on a gambling-type variable ratio, for example, or the reverse, to motivate yourself to continue to work despite ratio strain dragging you down. Some coping techniques were worked out long before schedules were, in the same way that farmers used selective breeding before scientists discovered genetics. An old tradition in the eighteenth century launched a long theatrical evening with a short one-act crowd-pleaser. Such little reinforcers can get us through the boring start of any big project.

Oddly, the feeling of victory afterward is often fleeting, followed by a letdown and a pause. That doesn't mean the consequence wasn't effective—clearly not, since you managed to finish the project. (We'll confirm repeatedly that emotions don't always align in the ways we would expect.) Humorist P. G. Wodehouse quipped that it was like feeling "somewhat filleted,"[22] and research verifies the anticlimax as a typical side effect when a lot of work or time is required for each reward. The slacking-off pauses at the hinge factory provide an illustration on a small scale. On a larger scale, a well-known bird-watcher tried for over fifteen years to find the rare and beautiful Connecticut warbler without success. Then—he sees the bird! "'I felt great for a minute or two. Happy, satisfied, relieved. . . . And suddenly I felt sad. I mean,'—he took a deep breath—'now that I've seen the Connecticut warbler, *what the hell am I going to chase for the rest of my life?'*"[23] But we can counteract these schedule effects too, for example, by having another project underway already, another goal to work toward—another consequence to keep us motivated.

SCHEDULES EVERYWHERE

You have to be on your toes to stay on top of schedules, though, because they routinely operate even when we don't realize it. Children in one school were given orange juice for a while to boost their nutrition. This intervention inadvertently doubled as a schedule of reinforcement for school attendance—which duly increased during this period![24] Where consequences are, schedules are too. Schedules are *everywhere*, even in our own minds. Ever struggle with a knotty personal problem at 3:00 a.m.? When persistently tugging at the angles might bring relief, it can be hard to stop obsessing. It can be yet another addictive variable reward schedule, insidiously working away.

Schedules have yet more unexpected powers. In one characteristic study, pigeons pecked for food on two different schedules. Then they were given a sedative. Their pecking slowed down as expected on one schedule, but simultaneously *increased* on the other.[25] So sedatives aren't just "downers," they can be "uppers" in this sense—depending on the schedule of consequences. No wonder consequences are critical to our understanding of nature-nurture and mind-body relations.

Schedules can even *determine* the very value of a consequence. Paradoxically, what is easier to get can become less valuable—as in colorful but weedy dandelions versus the boringly white but fiendishly hard-to-get ghost orchids referred to at the start of this chapter. There are plenty of examples; just watch *Antiques Roadshow* (PBS), or recall children learning to ration their favorite songs (see chapter 1). Sadly, this rarity effect inflates value, and can make it profitable to hunt rare plants and animals to extinction. On a happier note, great books pack a punch *because* their rewards require more thought and effort than the grocery-store potboiler, or the CliffsNotes version, for that matter.

One of my favorite examples of this paradox takes me back to my childhood. I used to enjoy fishing, and I'd spend hours watching a bobber drift to and fro, waiting for a nibble. So I was intrigued by a fish-in-a-barrel roadside attraction: you paid for the right to fish in tiny ponds stocked with hungry fish. And you were guaranteed success or your money back! Well, that may have been great for kids, but most adult a-fish-onados would be downright disgusted: how boringly easy to get a strike, and how unreinforcing these fish. You might as well buy them at the market. The difficulty of success of fishing in the wild (a lean and variable schedule of reinforcement) corresponds to the legitimate pride of achievement, rewarding patience and skill—and making it more appealing.[26] It's ironic that in the Western world we strive to make desired consequences easy—but what's easy can become worthless. And that's a schedule effect.

Consequences are almost infinitely varied and so are the schedules on which they come. How often is the consequence available? Under what circumstances? Does it depend more on the amount or type of a behavior? And what about timing? Maybe a signal has to be present, or a series of different behaviors. And the schedule can change, perhaps because of events beyond our control, or perhaps because of what we ourselves do. These changes themselves have reinforcing, punishing, and signaling effects. And there are always lots of other consequences available at the same time, and *they* affect what happens. There are even schedules of schedules (seriously)!

They all produce predictable patterns of behavior—although we'll see that those patterns can depend on our histories and whether we know what's required. And they apply not just to rewards but also to consequences we *don't* like, the unpleasant things in life.

Chapter 6

THE DARK SIDE OF CONSEQUENCES

Children often wish for the inability to feel pain, but those born with that condition seldom manage to live long enough to reach adulthood. Victims can bite off their own fingers without realizing, and one child needed both legs amputated by age eleven. Pain is as necessary as pleasure.

—Susan M. Schneider

Every day we cope with the other side of consequences. A child rushes outside to enjoy the first snow of winter—without a warm coat. Continued frolicking is punished by the freezing temperatures; a quick return inside is reinforced by escaping them. By definition, reinforcing consequences keep us going while negative, punishing consequences slow us down. Thus escaping a negative is a reinforcer, and so is avoiding it in the first place—slipping into a coat *before* venturing out.

That seems straightforward, suggesting a simple continuum of consequences from negatives to neutrals to positives. Avoiding negatives might seem to fall somewhere toward the middle, but you know by now that it's never that simple. Indeed, avoiding a dreaded negative can be a reward as strong as any dazzling shower of conventional positive reinforcers. Suppose you plan a dangerous operation because the consequences of *not* having surgery could be fatal. Negatives don't get much bigger than that. You prepare yourself grimly, study the procedure, say farewell to your loved ones, and make a will. At the nth hour, you decide to get a second opinion—and are assured you'll be fine without the surgery. Spectacular rejoicing! What can equal it?

Author William Least Heat Moon: "You never feel better than when you start feeling good after you've been feeling bad."[1]

At the other end of the spectrum, a single strong negative can inspire desperate efforts against a repetition. Some victims of crime move out of town at great personal sacrifice, for example. In other cases, a slew of opposing positive reinforcers can fail to counteract one early negative. The "dark side" of consequences is a force to be reckoned with.

None of us can escape that force: from death to taxes, from tornadoes to traffic jams, negatives big and small plague our lives. Montaigne wrote, "every hour I jostle against something or other that displeases me,"[2] and perhaps that's not just a literary exaggeration. Even turning your head when you're tired is distinctly annoying, or struggling with a plastic sandwich bag that won't zip. A pause of more than a few seconds in a conversation is unpleasant enough that many Americans will say almost anything to fill the gap, researchers find. (Cultures have different "rules" in this regard.[3]) Some folks simply develop thick skins. Consider Cindy, who walks into the house right by some tramped-in dirt clods without even noticing. Stan kicks the bigger ones outside and grabs the dustpan for the rest. Why do the neatniks in the family end up doing the cleaning? If it *doesn't* get done, it drives them wild, a consequence worth some effort to avoid.

It's an ill wind that blows nobody good, however: manufacturers take advantage of small negatives to boost their sales. I've noticed that flip-tops are becoming common on soup cans in the United States. Using a can opener must be enough of a bother to create demand. Paper towels in restrooms require some effort to dispense, a carefully engineered negative so people don't take more than they need.

SHADES OF GRAY

Then there's real work. Kids and parents alike, we can go to great lengths to avoid doing our chores. But what we *call* work isn't always a negative. Huskies pull sleds as eagerly as Babe Ruth played baseball, and most jobs have their compensations. Profound insight: even chores with no apparent appeal can be transformed into desirable rewards. For example, Tom Sawyer's cleverness made his friends fight for the privilege of painting his aunt's fence for him. And a negative can become a positive in the blink of an eye—or vice versa. Gossip transforms friends into enemies, and the good news about your promotion turns to ashes if a rival got a bigger promotion. Is that glass half-empty or half-full?

For some, the dark side takes over and it's *always* half-empty. Such people can live in fear, surrounded by negatives—past, present, and anticipated—their lives becoming a gauntlet of escape and avoidance. Animals can too, with abused pets being sad examples. Whenever even milder negatives like dog choke collars and reprimands predominate, in teaching or in everyday life, the dark side triumphs. In chapter 11, we will see research recommending a sort of "magic ratio" of about 5:1 for positives to negatives, for marriages and other human relations.

So are negatives bad and positives good? Not only is it not that simple (recall the kids who bite off their own fingers) but it can also be tough to tell the players apart. Distinguishing between working *toward* a positive and *away from* a negative is surprisingly difficult. Is dinner more an enjoyment of delicious food or an escape from hunger pangs? (How good of a cook are you?) Intentions don't count, as the families of poor cooks can testify—and so can teachers who offer "extra credit" points. Instead of viewing them as attractive bonus rewards, some students count on the extra credit and complain if they fail to qualify.[4] Are they happily working for a desired positive or fearfully avoiding the negative of missing out? Why not both in turn, or even simultaneously? Both provide effective motivation. It wasn't unexpected when research suggested that escaping a shock can have similar rewarding neurophysiological effects as receiving food.[5] And the schedules of consequences work similarly either way.[6]

I find hiking intrinsically rewarding, much like enjoying a sunset or watching a favorite movie. Because it can serve as a reward, I might wait to take a walk until I finish some work, in order to help keep myself motivated. But sometimes I hike when I'd rather not, because I need groceries, for example. Sometimes I hike both because it's enjoyable *and* because I need groceries. I hike because it's healthy (a long-term rule). I may also be avoiding work, although

I don't always admit that to myself. I may not always even be aware of it. (I do need groceries, after all.) The same behavior can have lots of consequences, lots of causes. My brother often jokes with me naturally—it's rewarding—but if he suspects that I'm stressed, he might do so more because he wants to cheer me up. The same consequence with the same effect on behavior can be given in the same situation for different reasons—different consequences for the *giver*. That's another part of the complexity, the mutual dance of consequences.

And everything is relative. If you have to choose one, going for groceries sure beats digging ditches; you'll even work for the opportunity (making it technically a reinforcer in that context). Suppose two young people are doing their chores so they can go out, which means they're on a reinforcement schedule. One kid's schedule is a lot easier than the other's, though—peeling two potatoes instead of a dozen, say. Then both kids are switched to the same intermediate schedule, peeling six. Which youngster will grumble? A rewarding improvement for one—an easier deal—is a negative for the other. It's like carrying twenty pounds of bricks after you've gotten used to carrying thirty—or fifteen. [7]

Animals respond to reinforcer relativity too. Again, it's not always straightforward, and context can shift what looks like a simple equation. One mild shock level had the effect of shutting down rats' lever presses for food. When other rats lever pressed only to avoid shock—no food involved—that same shock level was simply ignored. [8] Go figure.

FEELINGS

Feelings seem like good guides to positives and negatives, but they are not as reliable as we might think. Trying for a positive often brings pleasant anticipation, but ratio strain can quickly make that positive too hard to get: there's a limit to how many hours most of us will stand in line for a long-awaited concert or the chance to buy the latest exciting techno gadget. Meanwhile, oddly, working to avoid a negative can actually be enjoyable. Being given a deadline, for example, suddenly makes a reinforcer available for meeting it. (Thank goodness for such small mercies.) Of course, deadlines have a way of being backed up by sobering negative consequences.

A daily hassle for many of us, procrastination can be analyzed as a sort of dance between these opposing forces. As a deadline approaches, the negative pressure builds, but simultaneously so does the relief of finishing the chore. Chronic procrastinators know this well, and some thrive walking the tight-rope of these balancing drives. Others suffer but can't work up enough moti-vation to get on task (insufficient consequences). Then there are those who so dread looming deadlines that they compulsively finish projects long before they are due. No wonder that feelings don't align in a simple way. Two people on the same deadline can behave in precisely the same fashion to obtain relief from the same negative—but one is a miserable bundle of nerves, while the other thrills to the chase, so to speak.

Success likewise can bring a variety of emotions—and the nerve-wracked may actually enjoy the greater impact, as we saw with the earlier example of avoiding surgery. In one of my favorite animal stories, Sierra Club founder John Muir immortalized a friend's dog, called Stickeen. With Stickeen as an unwanted but eager companion, Muir crossed an unknown glacier, one that proved to be very dangerous. As dusk approached, he became compelled to pass over a long, deep ice crevasse spanned only by a sliver-bridge (shades of *Lord of the Rings*). Muir, pushed to his limits in getting himself across, was unable to help Stickeen. The dog showed how afraid he was to follow, but finally made the attempt after frantic coaxing. When Stickeen succeeded in traversing the chasm, he showed an ecstasy of relief. "Never before or since have I seen any-thing like so passionate a revulsion from the depths of despair to exultant, tri-umphant, uncontrollable joy."[9] After this drama, the dog became inseparable from Muir—his presence became a very powerful reinforcer. Never mind that it was his fault for leading them both into danger!

To approach this euphoria, some embrace extreme sports like skydiving—perhaps those same chronic procrastinators who thrill to the chase. What a contrast to those who avoid an aversive at any cost, such as those with strong phobias of spiders. But then, avoidance is notoriously hard to overcome. Successful avoidance, after all, means never learning whether the negative is still there or if it's as bad as expected. We might even like what we avoid. Indeed, reformed arachnophobes have been known to adopt pet tarantulas—maybe as a satisfying reminder of their victory over fear?

CHOOSING PAIN

Escaping a negative is one thing. But why would anyone freely choose to endure pain? Punishment by definition is avoided, after all, and pain tends to be an effective punisher.

This isn't really a paradox. Choosing short-term pain in the service of longer-term gain is common enough: better an injection now than a serious disease later, for example. Weighing consequences over time can change their value and their function.

We may come to enjoy mild hunger pangs, for example, if we control when to end them. But we do so because we have learned that they make food more reinforcing. ("Hunger is the best sauce.") In fact, we may deliberately let hunger pangs grow, negative as they are, to increase our dining pleasure later. Some of the ancient Romans famously took this principle to an extreme, voluntarily inducing vomiting midway through a long meal so they could continue to gorge themselves (although this story may be apocryphal).

This principle can be taken to still greater extremes: new gang members regularly choose to undergo an initiation of a beating. The consequences for *not* being in a gang can be severe, so teens "freely" undergo this ordeal.[10] Everything is relative, indeed.

Fortunately, negatives can be transformed in happier ways. Overwhelming them with positives is one strategy: Singing while you work has been popular for millennia, and chronic pain sufferers testify to the value of pleasant distractions. In a bit of laboratory legerdemain, an aversively loud noise became a rewarding choice when it was sugarcoated with enough positives. Given two knobs for participants to pull, the one that produced noise was avoided until the noise was made a signal for a rich reinforcement schedule.[11] The reinforcers compensated for the unpleasant noise. Even shock works, although either the shock has to be mild or the rewards it signals have to be fairly hefty.[12] I know that I will tolerate annoyances like cigarette smoking in a friend that I wouldn't accept in a casual acquaintance. If the friend is close enough, pleasant associations can even make the smell of smoke mildly rewarding in itself. If the relationship ends badly, the same odor can become aversive (see chapter 8).

Alternatively, negatives can lose their bite when we simply get used to them—technically, we become *habituated*. Simple repeated exposure to an aversive, even a moderate shock, can bring tolerance.[13] Human "pack rats" habituate to a level of clutter that astonishes others. Freeloading herons at fish farms eventually ignore any hazing technique used to frighten them because it becomes routine. Unfortunately, people also habituate in a real sense to violence and poverty suffered by others and stop caring—what's sometimes referred to as *psychic numbing*. Some negatives need to stay negative.

Boredom is an odd sort of negative featuring no reinforcers. It drives us to look out the window, turn on the radio, recall a happy memory—or even get a long-avoided chore out of the way. For lack of a more reinforcing alternative, doing the chore acquires sudden reward value—whether it's more a matter of escaping the pressure to do it or, more positively, appreciating it and whatever it accomplishes.

Clearly, boredom drives us to learn to make the most of what consequences are available to work toward. The changeability in their value to us is testimony to our ability to find rewards against the odds. Researchers of psychology's *adaptation-level phenomenon* discovered that even people who lose their sight or their mobility often adjust rather well, finding a new equilibrium at levels of reported happiness that are sometimes close to what they had previously.[14] It's reassuring.

AGGRESSION

Less reassuringly, the reaction to negatives can go beyond whining. A rat that gets a shock may strike out at an innocent bystander rat, and the same result has been found for other species.[15] One study found how history can make a difference: rats were less likely to attack each other after mild but unavoidable shocks when they'd had prior experience with unavoidable shocks.[16] Other research showed that it's not just pain that brings out the worst in animals. Transitions to poorer schedules can have the same effects as shocks, for example. Pigeons that had been rewarded on rich key-pecking schedules got violent when the schedules became leaner (providing fewer rewards) and more pecks went unreinforced.[17] Stress-hormone levels in rats went up when they were switched from rich to lean schedules.[18] (Similarly, pigeons choose to escape from difficult fixed work-based schedules during the lean post-reward period before they start a new run.[19] After all, they've learned that no rewards will be available during that time.) Ending a reinforcement schedule entirely—"schedule extinction"—is also likely to bring protests or worse.[20] The phenomenon is known as "aversive-induced aggression."

Obviously, people share the same susceptibility to aggression when negatives strike. In one lab study, high school students worked for money while periodically being annoyed by a loud tone. They could escape it by an easy button press *or* by a forceful blow to a cushion. When schedule extinction ended the earnings but the annoying tone continued, the cushion became a punching bag for most of the students, even though the easy button press still worked just as well.[21] Most of us can relate.

It's a bit harder to relate to this finding: given a negative—even a mild one like a period of schedule extinction—the chance to show aggression can be such a powerful reinforcer that animals will actively work for it when no innocent bystanders are handy. In one study, for example, pigeons on a fixed time-based schedule key pecked for food. They could also peck another key to get brief access to a target pigeon to peck at. (The research was set up so the target bird avoided the attacks by hiding behind a barrier.) The key-pecking pigeons frequently worked for the chance to attack during that unpleasant period right after the food delivery, when there was a long way to go until the next one.[22] It's like an aggrieved boss calling a random worker over to swear at. Fortunately, the aggression object doesn't have to be a fellow creature. It can be innocuous, such as a bar for a rat to bite on. Who hasn't slammed a door in anger?

You would think that with all the negatives surrounding us, people would see the need to go easy on one another. Easier said than done! Unfortunately, giving out negatives can be as reinforcing as escaping them. In a vicious cycle, it can *mean* escaping them. Annoy me, and how natural for me to annoy you back to get you to stop. Sometimes it works—it is one of those addictive variable-reward schedules. Brothers and sisters are experts at this game. When parents scold in turn and the kids shut up, the parents have just been immediately rewarded. No wonder so many of us wham each other with reprimands and criticism more readily than we do with praise. (So much for the magic 5:1 rule in favor of positives.)

MAKING NEGATIVES WORK—POSITIVELY

Sometimes punishment *is* the best choice and there's no getting around it. For a behavior that needs to stop right away, like a toddler running into the street, a consequence strong enough to do the job immediately is best. As in the case of the begging dog in the previous chapter, researchers find that what may seem more merciful—using weak negatives first—can end up being cruel. If the behavior recurs, you may have to escalate to a stronger punisher to stop the behavior than if you'd started out with a moderate one.[23] That said, progress in our understanding means alternative strategies are usually available.

That's fortunate for a number of reasons, one being the difficulty of getting aversives to work as intended. Animal-training expert Karen Pryor related the story of a boxer dog who loved to lie on a forbidden couch when his human was away. Confronted with mousetraps on his favorite spot, the dog dragged a blanket over the traps, setting them off, and then lounged at ease.[24] If there's a way to avoid a negative, it will be found.

It is also fortunate because negatives carry a number of problems and unpleasant side effects, such as the avoidance and aggression we've been discussing.[25] One of the most important is that negatives don't teach what is desirable. Alternative approaches involve positive reinforcement, which is much more fun all around *and* better in other ways, too. Alternatives will be discussed at the end of this chapter and in later chapters.

Meanwhile, how do we best handle punishment when it *has* to be used? A schedule that manages to provide a punisher for every occurrence of the problem behavior will stop it most quickly. Under these circumstances, the only behaviors that continue are strongly rewarding in other ways. Intermittent punishers are naturally less effective, although often they are all that's possible. Just as for positives, variable intermittent schedules should work better than comparable fixed ones: it's harder to keep going if you know the very next errant behavior might bring retribution.

For the same reason in reverse, it can be hard to stop using aversives once you start. Even when they don't work as well as hoped, occasionally you succeed, so using them can itself be reinforced on an addictive variable schedule. Using aversives also offers a chance at control, which we know to be a powerful reinforcer (see chapter 1).

The temptation to use punishment is understandable, then, and so is the temptation to indulge in aggression when you have been punished. How, then, can we stop the aggression? The knee-jerk reaction of more aggression can work—punishment in action—so it's not surprising that it is so common. Ironically, animal research has shown that the same shocks that cause aggression can stop it, too, if delivered as an immediate consequence for the aggression. Even shocks at a lower intensity can work.[26] But again, other methods are usually preferable (see chapters 11 and 13).

Doing nothing and hoping for the best would seem an easy way out, but it has consequences as well. (Everything does! We can't escape consequences!) Aggression—or other undesirable behaviors maintained by other rewards— will simply continue if we look the other way. Ignorance is seldom bliss. (The "ignore it" approach is bad for desirable behaviors as well: they can inconspicuously ebb away without some sort of maintaining reinforcer like our attention. What might happen if a child is never praised for sharing a particularly valued toy?)

Fortunately, there are a ton of options. Providing a positively reinforced alternative to any undesirable behavior should be the rule these days rather than the exception. How can we deal with aggression? One group of researchers tried rewarding nonaggressive play in grade schoolchildren with raffle tickets. Kicking and hitting declined by large amounts.[27] (Ideally, the natural reinforcers of more peaceful play would then take over.) Similarly, when a dominant male chimpanzee's aggression was a problem during meals, he was made the focus of attention and food rewards in a separate area. Friction plummeted as the other chimps got to eat in peace.[28]

"Best practice" still includes a place for negatives. Common examples include mild ones like reprimands and "response cost" (for example, a fine— the removal of a positive such as money is a negative). Reprimands can be very effective in stopping a behavior, and, surprisingly, soft ones sometimes work best (as discussed in chapter 14). Timeout from positive reinforcement, a mild negative formally introduced by experts in consequence principles in the 1960s,[29] has become the preferred alternative to spanking for many parents. Keeping kids in at recess—a form of timeout—has produced clear reductions in aggression. (But why not use the positive-reinforcer approach instead or in addition?) Other strategies discussed later in the book include altering the problematic circumstances in a variety of ways, and providing rules and reasoning. The best approach of all is to change the motivation in the first place— the value of the maintaining consequences *and* the alternatives. Then it's the healthy, desirable consequences that are both more rewarding and available. The next chapter helps put consequences in context.

Negatives can be downers, there's no escaping that. But we've seen how lifesaving they can be—how grateful we should feel for evolution's painful solution. And let's not forget that positives have a negative side, even when good feelings abound. The health problems of obesity and smoking show that we often choose pleasure now despite deferred pain. Sometimes it's a shame that delayed negatives *don't* have more of an effect. Another unexpected virtue of most negatives is the obvious visibility of their dangers. The dangers of positives are not only harder to see, they're harder to combat.[30] What truly is the "dark side" of consequences?

Chapter 7

CHOICES AND SIGNALS

"Two roads diverged in a wood, and I—
I took the one less traveled by."
—Robert Frost, "The Road Not Taken," 1916

Frost's poem captures the imagination, and what schoolchild in the United States has not pondered its meaning? Teachers naturally focus on the consequences of unpopular, less-traveled-by decisions, and our life-changing choices shimmer in bold relief.

But the truth is that we are immersed in choices, just as we are immersed in consequences—in fact, the choices exist *because* we are immersed in consequences. Any time we decide what we "feel like" doing, for example, we are really weighing consequences.

We hardly go two minutes without making a choice. On the job, is it time to study a spreadsheet, consult a coworker, or plan for the weekend? Or maybe just pick at a fingernail. We have choices even when we're thinking to ourselves: it may be fun to daydream, but given a more immediate problem to solve, we choose to switch gears.

Back at home, should we watch sports or a sitcom? And what are we going to eat? Large US supermarkets now stock *40,000 items*,[1] overwhelming even European visitors. In the developed world generally, we now have so many choices that some economists and psychologists have argued it's a liability. If so, it's one we don't have the choice to escape.

There's no lack of help, anyway: our friends, our relatives, and a huge number of books advise us on how to make better choices. That's every single dieting bestseller, plus all the manuals on the self-help, parenting, and consumer shelves. But dealing with choice means dealing with consequences.

THE MATCHING GAME

Because choice is so ubiquitous and important, hundreds of studies have examined how it works. Let's begin with some human research. A simple nod or grunt can reward someone for starting a topic of conversation—this is not surprising. In one study utilizing mildly supportive grunts as consequences, college students were asked to recall memories of their childhood. After they had reminisced freely for a while, half of them were rewarded in this unobtrusive way for describing family memories, and the other half for nonfamily memories. More than 80 percent of the students modified their commentary accordingly. And—another theme of this book—most of the students failed to realize what was happening, so unaware are we of subtle consequences (see chapter 9).[2]

Now suppose that you're sitting at a "mingle" table in a cafeteria with two strangers. One of them nods, smiles, and affirms your comments more than the other. Who will you talk to more? That seems pretty obvious. In one study, people ended up spending about twice as much time talking to the rewarding companion as to the wet blanket. As it happened, they also received about twice as many supports from the rewarder. Their conversation choices "matched" their conversation rewards, with a 2:1 ratio for each.[3] This turns out to be a characteristic result when a time-based component is involved and when a series of rewards is available, not just one or two.

For example, suppose you and your significant other are surfing your eighty-channel big-screen TV. Your SO wants to watch a nature show with spectacular big-cat attacks. You want baseball. You switch back and forth between the two, attempting to maximize your viewing pleasure. If the baseball game is kind of slow, not much happening, you'll spend a lot of time watching leaping lions. But you'll keep switching, and the amount of time you spend on each channel should (in theory) come close to matching the number of rewarding sequences. In the same way, multitaskers can deftly skip back and forth between two electronic devices and one real person in the space of one minute.

The *matching law* was originally derived from animal research in the lab, where conditions can be precise. In its full technical form, the equation gets complicated, covering a host of factors and parameters: bias between the behavior choices (an SO who really dislikes baseball), different levels of effort (someone lost the remote, so you have to get up and change channels manually), different types and values of the consequences, delays, schedules, signals, and so forth.[4] Some of the current mathematical models are so high-tech that they employ the differential equations of calculus.[5] Further, in real life, we can be downright inundated with choices, and my own research has helped demonstrate matching with up to four options. Other studies have confirmed even more.[6]

Then there's the necessity of dealing with negatives. A choice that comes with a downside loses some value (the baseball game has commercials, the nature show doesn't), but how do the positives and negatives combine? Sometimes straightforwardly, sometimes less so, it turns out. We are only dimly aware of some consequences, and some of course are substantially delayed, reducing their value—say you want to be able to talk over the game with your pals over the weekend (see chapter 12). All this is not easy to disentangle. What's important is that even the simpler forms of matching appear to offer a decent description of comparable choices in real life.

The surprise isn't so much that choices track consequences in some way. (Imagine how haywire things would be if they didn't.) What was eye-opening were the mathematically consistent and elegant ways in which they did so across so many species, consequences, and behaviors. From domestic animals like cows and chickens to wild species like bluegills, brushtail possums, and coyotes, animals have shown their decision-making skills and "followed the money," just like people.[7] Consequences have varied from social reinforcers to "pleasure center" brain stimulation, from cartoons to drugs; also money, warmth, and (of course) edibles of all kinds.[8] Scenarios have included wild birds foraging in their native habitats, students working on arithmetic problems, and even human conversation, as we've seen. All have produced orderly patterns of repeated choices, well fit by the matching law equation.[9]

Birds in their native habitats? One careful study investigated birds foraging in the wild for insects and spiders. Most of these birds (called "wagtails" for good reason) stayed with a main flock that roamed a large prairie with a fairly constant supply of bugs. But some wagtails had their own territories where these tasty items washed up on a riverbank; the longer between visits, the more food that would accumulate. These birds had a choice between feeding with the main flock or going off on their own. The time they spent in the two areas followed the matching law.

Something like the matching law is obeyed by groups: the richer the reinforcer area, the more individuals that will naturally be there. Even fish have been shown to follow a matching-type distribution across foraging patches of different value.[10] Of course, other consequences must be considered, too. One innovative researcher showed how social dominance (the "pecking order") and flocking for protection affected matching for food by ordinary house sparrows.[11] (Dominant birds did better than subordinates, not surprisingly.)

One recent computer model was focused on mimicking selectionism, "the cause that works backward" in evolution and in learning from consequences (variability, then consequences, and then the resulting change in the repertoire). With natural behavioral variability and the history of consequences included, the mathematical model reproduced typical behavioral patterns in choice (matching) and in simpler variable work-based schedules.[12]

SO WHAT CAN THE MATCHING LAW DO?

The matching law can actually *quantify* comparative value. That is useful in all sorts of ways. Suppose you, a farmer, want to find out how much your cows prefer expensive barley over cheaper options. Researchers like New Zealand's Mary Foster have used the matching law to discover exactly how much cattle preferred their favorites over alternatives like chopped hay, ryegrass, and kiwifruit—and how these culinary selections compared to one another. The resulting precise preference scale described most individual data sets well. A few cows with nonconformist tastes had their own scales. (Someone always has to be different.) Once compared to one of the standards, a new feed could be fit directly into the scale: testing confirmed that its relationship to the other feeds could be predicted on that basis, a useful feature called transitivity.[13] These results are valuable for farmers trying to balance cost, nutrition, and acceptability to their herds.

Another study quantified the degree of dislike hens had for loud sounds.[14] Factory-farm noise levels can reach 100 decibels, enough to cause hearing damage in people. One of the most unusual of such animal-welfare studies had pigs choosing between straw, fir branches, or peat as rooting materials.[15] Given these examples, it shouldn't be surprising that dog-food preferences are sometimes assessed in the same precise way.[16] That's big business, after all.

Indeed, Wall Street hasn't ignored the implications: matching can occur when people shop on the Internet, for example (click, click).[17] It's also been shown to apply to heroin addicts giving themselves the therapeutic drug methadone.[18] The matching law may even work for football quarterbacks calling plays: the more yardage gained for passing instead of rushing, the more attempted passes. (However, a revealed bias against passing reflected the greater risk of turnovers and more variability in gains.[19])

In the lab, people often match in the same way that animals do, but not always.[20] These exceptions are informative: With weak reinforcers or low-effort behaviors like keyboard presses, participants sometimes say, in effect, why bother to track the rewards carefully? It's like having the TV on just for the sake of having it on. You're not paying much attention, so even though you're surfing the channels, you don't choose carefully based on what the reward value would more normally be. Then there are instructions and rules:

ironically, they can help us be even more efficient or, if they're wrong, throw us off entirely (see chapter 10). Both of these effects impacted one human matching study in which several participants decided that the payoffs were roughly equal for the different choices, even though that was not actually the case. From then on, they responded equally or randomly to each choice, losing out as a result. Ironically, animals performed significantly better, as did other participants who proved to be more sensitive to the relationships.[21]

WINNING MATCHES

Basic research on choice schedules has turned up some unexpected findings with practical applications for people. In several of my own studies, for example, I found that the most frequently reinforced choice often occurred even if a different behavior had just been rewarded.[22] If you're trying to break a bad habit by encouraging a better but less appealing alternative, this finding can help you understand why the change seldom happens overnight. An off-task schoolgirl just praised for a spell of diligence might revert ten minutes later to chatting with a classmate. Consider the choices and their consequences (and consult chapter 14).

Other human applications are widespread. Kids worked on a computer so they could take a spin on an exercycle or watch a video. One schedule of reinforcement applied to the exercycle, and a second to the video, and the children could switch between the schedules freely. The children's choices obeyed the matching law, and were predictably influenced by the price to participate—in this case, how hard they had to work on the computer on the two schedules. When the schedules were changed so that working for video watching was harder, most children switched to the exercycle. The most obese children were less influenced by this schedule change, continuing to show a strong preference for the sedentary activity.[23]

Conflict in prekindergarten followed a matching-type relation too. Aggressive four-year-old boys turned out to be successful using yelling and hitting at home. In their families, these methods worked to get what they wanted, while nonaggressive methods were less likely to be rewarded. Boys with more acceptable ways of coping had families in which the nonaggressive methods were more likely to work.[24]

In their bestselling book *Nudge*, Richard Thaler and Cass Sunstein argue that we are all "choice architects," weighting consequences for others as well as for ourselves. They used the example of food displays in cafeterias that foster healthy or unhealthy choices through changes in factors like behavior effort (what's easy to reach?) and visibility. Someone might want that salad, but if it's on the back row, its reinforcing value might be insufficient to offset the effort of getting it. (There's a reason that manufacturers pay for better positioning of their products in grocery stores.) How much better it would be, the authors suggested, if the understanding and planning implied by "architect" went with the job.[25]

In the previous chapter, schoolkids at recess were rewarded for playing nonaggressively. Those kids had a choice. To reduce a problem behavior like aggression, you don't always have to focus on it; you can instead provide a healthy alternative with a greater reinforcer value. That's being a choice architect. Likewise, two researchers looked at the bigger picture—the context—and found that success in weight loss and alcoholism recovery was sometimes linked with new alternative reinforcers in people's lives such as new friends.[26]

So, context matters a lot. Delayed and immediate, large and small, there are always other choices, other consequences. And there are usually signals that go with them.

GETTING THE SIGNAL

When the same action can bring punishment in one setting and lavish reinforcement in another, it helps to be able to tell the difference. Signals for consequences let us do that. These cues let us all pick up on when to stop and when to go—literally, in the case of traffic lights. The links to consequences are clear. Kids quickly learn that they had better not swear in the presence of their parents or teachers. With their friends, it's a different story, and there are clear links here too. When the boss is frowning, forget about asking for a raise. See a police car? Check your speed. (Taking advantage of signals, some precincts post cars with mannequins in them to utilize their staff more efficiently.)

Animals get the idea too; remember the little lever-pressing bee in chapter 2, switching levers in response to a signal light? A famous naturalist pounded suet into holes in a tree trunk. The local woodpeckers loved it so much that they learned to fly over when they heard the hammering.[27] Similarly, a wild mockingbird started coming for food in response to a whistle, even approaching its benefactor and imitating the whistle.[28] (Mockers imitate everything, so not too much should be read into this.) It's recently been established that wild mockingbirds quickly learn to recognize individual people of importance to them.[29] This might make us look at *them* differently.

Dogs read our signals—including scents—to know when we're thinking of taking a walk that might include them. (Sometimes they know before we do.) Subtle cues present a challenge, but most of us become reasonably good at reading expressions and emotions, particularly of the people and animals most important to us. The consequences that go with them are especially important, after all. These skills are invaluable to social species like the dog's ancestor, the wolf.

Signals get crossed or missed all the time, of course, and there is a whole science of how they work—or don't work.[30] To be effective, signals have to be noticed and they have to reliably predict consequences. That sounds straightforward, but specialists in ergonomics can testify that it's not. Optimizing the design of airplane control panels, for example, was a major challenge for many years. Even everyday signals can be counterintuitive. Feeling tired can cue a need for exercise, odd as that seems. Science is one of the best ways of establishing reliable predictions about consequences and cues. What weather patterns grow hurricanes? Science has told us.

We start learning signals when we are young: two-month-old babies accurately learned to turn their heads right when shown a checkerboard and left for a circle.[31] (The reward was a smile and praise, sometimes a toy.) Many newborn animals can learn signals associated with consequences.[32] Even primitive invertebrates like planaria (those tiny flatworms) readily pick up on arbitrary cues, as we saw in chapter 1. Honeybees could learn which of two patterns signaled nectar.[33] More complex features are no problem, at least for vertebrates: as we also saw in chapter 1, pigeons can not only tell the difference between contrasting musical styles, they categorize novel excerpts in the same way that people do.

Signals in the wild can be surprising as well. Some prairie and savannah birds have learned to seek fires for the easy pickings of available prey that fires signal, which explains the unexpected images of hawks and flycatchers winging *toward* smoke and flames.[34] Meanwhile, gazelles on those savannahs perform a leap that goes nowhere, called stotting, when they see a lion. Lions learn to avoid chasing high-stotting gazelles, who effectively signal what a waste of effort it would be.[35] (Consequences in action.) The higher the leap, the more fit the gazelle, and thus the more reluctant the lion to pursue.

Sometimes signals are "absolute," such as the perfect color for a ripe huckleberry. Sometimes they are "relational," like the biggest berry, the most dazzling peacock finery—and the brightest star to wish upon.

A SMORGASBORD OF SIGNALS

Back to earth, signals work for negative consequences, too, like the gazelles' message for the lions. ("Don't waste your time.") Being able to predict a negative means improving your chances of avoiding it, clearly a useful survival skill. Ants, honeybees, and crabs can learn signals for aversives as well as for positives.[36]

Reminiscent of the mockingbirds, gulls came to view a research ornithologist with suspicion, ignoring other people but scramming when they spotted their favorite villain. Attempting to outsmart them, the ornithologist wore a disguise, but the gulls soon saw through it. He had to wear a different outfit every day in order to continue his observations.[37] Similarly, nesting ospreys (fish hawks) were unperturbed at the sight of the owner of a boat moored nearby—*unless* he was wearing boots. Then they screamed and strafed: boots were a signal that the boat would soon intrude on their nest.[38]

A negative can *be* a signal, as we saw in chapter 6. In one lab study with pigeons, without shocks, no food was available. The shocks signaled that food was available for key pecking on a variable reinforcement schedule, so removing the shock turned out to produce a *decrease* in behavior.[39] That's like valuing a mean babysitter because she often took you to your favorite playground, one your parents never had time for. You would be sad when she was dismissed (to your parents' puzzlement). More often, of course, a negative signals other negatives—like the boss's frown signaling stormy weather ahead.

Signals can be the *absence* of something as well as the presence. Drab birds can be identified in the field from their very lack of patterning ("no wingbars" can clinch an ID). More powerfully, think of rats pressing a lever to avoid a mild shock. Each press earns a five-second period when no shocks ever occur, and the rats learn this "safety signal."[40] In a similar way, chronic nail-biters wash their hands frequently, producing a safe period when they can indulge their habit without fear of germs. The simple passage of time is a signal.

Our own behavior can also be a signal. In the simplest case, if we are washing our hands and someone asks what we're doing, we can tell them. Rats can do this too by pressing levers, accurately reporting behaviors like face washing and walking, even the "behavior" of not moving.[41] It sounds unlikely, doesn't it? Pressing a left lever might say "I was face washing"; a center lever, "I was walking"; and a right lever, "I wasn't moving." When a signal is given, the animals report what they are doing and get rewarded if they report accurately. (Rats are capable of pretty impressive stunts.) Sometimes we humans do things automatically, notice that we are doing them, and change our actions as a result (for example, oops, I'm trying to stop nail-biting). Observing what we're doing is a behavior in itself, one that can also be influenced by consequences (see chapter 9).

Given these results, it's not surprising that internal cues work fine. If I feel a tummy ache starting, I grab my favorite remedy and avoid the worst of it. The activity of new drugs is sometimes tested in an intriguing way that relies on such cues. Animals like rats (again) learn to press the left lever when they have been given one drug and the right for another. Then they are given a different drug. Will they press more often on the left or the right? If one choice predominates, the new drug has something more in common with that one than with the alternative.[42] Studies like this help researchers evaluate the drug's likely effects before testing it on people.

OF SIGNAL IMPORTANCE

I never thought I would appreciate swarming, buzzing Midwestern mosquitoes until I moved to south Florida. Suddenly I fell victim to stealth attacks from the silent version of the bug. I missed getting the warning.

We sometimes prefer signaled negatives even when they are unavoidable. I guess we want to know what's in store if bad things are coming. Animals can share this view: In one matching-law-type study, rats chose signaled (mild) shocks even when they were several times longer than unsignaled shocks.[43] These sorts of results are helping clinical researchers improve the treatment of anxiety disorders in humans. The choice between signals and no signals is not straightforward, though: seemingly contradictory (but not really), if the negatives might *not* be coming, then it turns out we often *don't* want to know about them in advance—the "ignorance is bliss" approach. In other words, if there is a chance we might get off scot-free, we'd rather not have prior notice that bad things might be looming. And there are other interacting factors as well.[44]

The same apparent contrariness applies when signals work in what seem to be completely contradictory ways. In one study, a chimp slowly worked through 4,000 lever presses for one (large) reinforcer. A human reading a rather boring 800-page book might feel like the chimp, but at least the page numbers signal progress, and finishing a chapter is reinforcing in itself. Adding a few progress markers of this sort to the chimp's task brought a huge increase in speed.[45] But this method isn't foolproof. For pigeons on a work-based fixed schedule with 180 pecks required, four added signals became discouraging reminders that there was still a long way to go, and the birds slowed down![46] How do we predict which effect will occur? It turns out to depend on factors like the particular schedule, the individual's own history, and the nature of the signals.

Many human tasks are lengthy, and using marker signals to break them up is routine. Those smaller, more manageable subsections usually bring built-in reinforcers. But again, there's a science to it. One application: Donors and volunteers alike appreciate fundraising graphics that track proximity to the goal— but mainly when significant progress has been made.[47] Relatedly, in "county listing," some bird-watchers try to find as many species as they can in each of their nearby counties. In one California system, rewarding signals of progress consist of color changes when each milestone of twenty-five is reached. It is so easy to find 100 species, though, that the color changes are rationed and don't start until after that point, when the real accomplishments begin.[48]

Any job that is long and tedious can benefit from this sort of analysis and application. The discoverer of the basic principles of consequences, B. F. Skinner, frequently applied them to himself. To boost his motivation, he kept track of how much time he spent writing each day and maintained a chart.[49] Watching the hours accumulate helped keep him going even on long projects.

Many behaviors are actually behavior chains: series of behaviors, each providing signals that lead to the next step. Rats can master impressively long chains: one star, called Barnabus, climbed a staircase, pushed a bridge into position and crossed it, climbed a ladder, pulled on a chain, moved a "car" across another bridge, went up another staircase, ran through a tube, and then descended in an elevator. At the end, he got food pellets, and then waited for the signal to do it all over again.[50] Such long chains are often best taught backward, as this one was. Start with the end, then the link before the end, gradually working toward the start of the chain but going to the end each time. Using this method, successful completion of a new intermediate behavior brings the chance to do well-learned behaviors that are closer to ultimate reinforcement. Along the way, these "getting closer" signals for consequences can become effective consequences themselves. (When the same scientist who taught Barnabus taught American Sign Language years later, he found that this "backward chaining" approach was the fastest way to teach the alphabet.)

Knowing about signals means we can use them to our benefit in many ways—including turning them off. Dieters avoid watching commercials for rich food, and smokers steer clear of smoking hangouts. Because e-mail sacrifices the nonverbal signals that are so important in everyday communication, smileys and other "emoticons" have become standard to help fine-tune the intended message. Even face-to-face conversation isn't exempt; it too benefits from nonverbal cues. She says she wants to go with you, but she's not looking at you, for example. One researcher has noted that when nonverbal cues contradict the words, we believe the cues.[51] We learn to detect which signals are actually good predictors of consequences. It's important.

Signals are so powerful that they often take on a hidden life of their own, as we'll see in the next chapter.

Chapter 8

PAVLOV AND CONSEQUENCES: AN ESSENTIAL PARTNERSHIP

*In February 1984, I returned from service overseas in the Peace
Corps. That August was the Olympics, and, like millions of others,
my family sat around the TV and watched the opening ceremony.
Each nation's representatives marched in, bearing their flag.
When the small contingent from my host country, Fiji, appeared,
I experienced a sudden rush of joy and nostalgia. I found myself
springing up exuberantly and shouting "Fiji!" (My family took
this in stride.) Even now, so many years later, the thought of that
episode elicits a recognizable remnant of deep emotion.*

—Susan M. Schneider

Feel a "gut reaction" like this and it's likely that you're experiencing Pavlov's learning process, "classical conditioning." Far from being limited to the famous salivating dogs, it is as ubiquitous as consequences are. Classical conditioning is built around reflexes and similar reactions that we don't have to learn, and there turn out to be a lot of them. Emotions and drug tolerance are on the list as well as eyeblinks, knee jerks, and startle responses. Some, like immune responses, weren't discovered to be classically conditionable for years.

Like learning from consequences, classical conditioning sounds a lot simpler than it is. Still, the basics are straightforward enough. Pavlov knew that dogs drooled when given meat. Given a signal like an electric bell shortly before the meat (and not at other times), the bell alone came to elicit salivation. It's efficient: after hearing the bell, Pavlov's dogs were ready to fall on their food. Like learning from consequences, this ability developed early in the history of life. Believe it or not, cockroaches salivate too and respond to classical conditioning.[1] (I don't know that anyone's tried a bell.) Neuroscientists have shown that, just as the principles are different, the neurophysiological bases for learning from consequences and classical conditioning are different.[2] Another difference: classical conditioning is based on a straightforward domino-type cause-and-effect relation, unlike learning from consequences with its selectionism ("the cause that works backward," with success reproduced later while failure isn't).

When I teach the college course called Learning, my students get to experience classical conditioning directly. We all have cups with lemonade crystals in front of us, and we keep plastic spoons with a dab of the crystals poised and ready. When I say the word "Pavlov," we put powder to tongue—and salivate, just like Pavlov's dogs. After repeated pairings, we do a blank, unpaired trial, and simply hear "Pavlov." The sensation of elicited salivation makes for a unique experience. Even though I know what's coming, I find myself smiling in astonishment, just like my students. (Try this at home.)

Hearing "Pavlov" repeatedly without any lemonade powder makes the word lose its power. Once, one of my students started being bothered by the taste and did not undergo this "extinction" procedure at the end of the session. Three *months* later, one of her friends asked about the class and innocently used the word "Pavlov." Bingo! A surge of salivation showed how long this form of learning can last.[3]

Salivation was the first classically conditioned response to be studied in depth; immune responses are among the most recent. Do they really follow similar principles? In a controlled experiment, rats allergic to egg whites were given that allergen while a complex light-and-sound stimulus played. Eventually, just seeing and hearing those lights and sounds activated their immune systems.[4] In another experiment, guinea pigs were made allergic to a protein found in cattle, and then an odor was associated with the protein. Presentation of the odor alone produced the same level of allergy-fighting histamines as the problem protein itself.[5] The effect has been confirmed in people, and our immune responses can be either enhanced or suppressed in this way. In one study, a drug that increases our immune system's "natural killer cell" count (a type of white blood cell) was paired with sherbet. The sherbet alone then elicited an increase in the participants' killer cells.[6] These sorts of results are now beginning to be applied clinically, for example, in fighting allergies.

COMPENSATING REACTIONS AND DRUG TOLERANCE

Also offering clinical applications are a special category of Pavlovian reactions.

Have you ever noticed that looking at food seems to make you hungry, and therefore makes food a more powerful reward? Classical conditioning helps explain this reaction. When you eat, your blood-sugar level rises. Studies show that the signals for approaching mealtimes can start eliciting compensating reactions, pushing your insulin level up and (as a result) your sugar level down—helping your body stay on a more even keel.[7] Because the sight of food is also associated with eating food, under some circumstances your insulin level can also become classically conditioned. Looking at food then really does change your body chemistry, and that in turn makes you feel hungry. (This is not a problem for grocery employees, because the association between seeing food and eating it disappears with extended experience.)

The linked signals for this classically conditioned insulin response can also be arbitrary, like Pavlov's canine chimes for salivation. For example, some rats sniffed an unusual nonfood odor that was paired with mealtime. Other rats got the smell too, but it was not paired with food. Later, only for the rats in the paired group, the smell alone was nearly as effective at jump-starting insulin flow as was a real meal.[8]

What gets classically conditioned, then, can sometimes be a response that compensates for the upcoming event, preparing the body much as salivation does. Compensating reactions can also be provoked by drugs (legal or illegal), and they have proven to be of grave importance.

Patients on chemotherapy sometimes get their injections at the same time each week, in the same room, even by the same nurse. That means ideal conditions for classically conditioned compensating reactions that fight the effects of the drug, trying (again) to keep body chemistry relatively constant.[9] Changing the room, changing the time, and having the injection given by a different person would make the drug work better because predictable conditions make the drug work less well. The result of predictable conditions can be drug tolerance, leading to ever higher doses with more side effects. Ultimately, the drug may become ineffective. Switching rooms and times can help delay the onset of tolerance. The same drug dose can demonstrably have different effects just depending on whether people inject themselves or are injected by someone else—it's evidently a sufficiently different experience.[10]

Tragically, some drug overdose deaths appear to be the result of this effect. In a familiar setting, you as a drug user have to take a larger dose to get a high because you have to override the compensating reactions that are produced by all the associated features there. (Note the corresponding change in the drug's reward value.) If you depart from your normal routine and take your usual dose in a different enough setting, the compensating reactions don't get elicited, and your system can get dangerously overloaded. Research with animals has confirmed the effect, and supporting evidence includes many human cases.[11]

In one of them, a man may have inadvertently killed his own father. His father, who suffered from pancreatic cancer, was being cared for at home in his dimly lit bedroom. He was in pain and received four morphine injections per day. On the day of the death, the son discovered his father in the brightly lit living room, where he had for some reason managed to move himself. Because it was time for the injection and his father was in pain, the son gave him his usual dose of morphine there. The father's reaction was exceptional, and a physician immediately diagnosed morphine overdose, but nothing could be done.[12] Although the involvement of classical conditioning can't be definitively established in a case study like this, discussion of its likelihood with an expert several years later helped alleviate the son's lasting confusion and guilt.

NOT ALL IN YOUR HEAD:
THE PLACEBO EFFECT
AND OTHER MIND-BODY SURPRISES

The placebo effect used to be considered an inexplicable, almost magical thing. Now, the same classical conditioning that helps cause drug overdoses also helps explain how sugar pills can make a sick person feel better. Really.

In one well-known study using fMRI, volunteers received shocks, but some participants were given a fake pain-relieving cream that they thought was genuine. The placebo effect occurred: these participants reported less pain when they used the cream. What was happening in their brains? Anticipation effects were evident as the shocks approached—but the "pain centers" were less active when the cream was used. The strength of the effect paralleled the amount of relief people reported, and that's a typical result.[13] No wonder then that placebos can become powerful reinforcers: they actually do help relieve experienced pain.

The phenomenon is not limited to people. Just as animals can develop drug tolerance, they can also experience the placebo effect. Associate cues such as colored lights and different flooring with effective drugs, then present the cues alone: animals like mice have been shown to receive pain relief *neurologically* as well as behaviorally.[14]

This does not mean what is involved is identical in both people and animals. For people, the placebo effect is partly a function of past consequences as well as classical conditioning. Taking what you think is a drug leads you to interpret what follows based on your experience with helpful remedies (see chapter 10).

Until these effects were widely acknowledged, the term "psychosomatic" was a dismissive, pejorative term. Now some healthcare experts have even suggested placebos as a treatment in themselves. Mind-body interactions are real, classical conditioning is part of them, and it's a revolution. It's not "all in your head," at least not in the way we used to think.

Simply viewing drug paraphernalia can activate the brain and make a cocaine addict feel pleasure or craving.[15] Going one step further, we now know that just *thinking* about something can have effects like actually experiencing it. No wonder daydreaming is fun; no wonder method acting works for performers.

Imagining dinner can cause salivation (try this when you're hungry).[16] Afraid of heights? Pondering your fear can hurt.[17] Similarly, conjuring up a spider may make you shiver—or smile, if you're a bug-lover like my brother. Thinking can elicit involuntary movements, heart-rate changes, and sweating, as well as pleasure-center and pain-center activation. Realize hours later that you have inadvertently eaten a cockroach, and only then do you get sick.

It gets pretty weird. An anonymous scientist famously noted in the prestigious journal *Nature* that when he had romance to look forward to, his beard grew faster. As a good scientist, he gathered systematic data to check his casual observations. Only before weekends when he got to visit his girlfriend did the effect occur, and this wasn't just time-based conditioning because it wasn't every weekend.[18] It turns out that the hormones affecting beard growth are subject to classical conditioning. The effect has been confirmed in rats: when an odor was paired with a randy gal, classical conditioning meant that the odor alone came to trigger as much testosterone secretion as the presence of the eager female.[19] Isn't it romantic.

GETTING EMOTIONAL

For many of us, just holding a souvenir from a long-ago vacation can recapture some of the happiness. No wonder overpriced tourist knickknacks sell: we know their value is more than appearances suggest. Likewise, digging out old family photos may be rewarding because of the gut reaction that comes from their pleasant associations—classically conditioned ones. B. F. Skinner described his wistfulness when he looked toward the site of an old familiar clock.[20] It wasn't even there any longer, but it didn't have to be. The context and his memories were sufficient. Emotions can be consequences in themselves, and because of classical conditioning, this potentially means anything that has been paired sufficiently with a strong emotion.

Of course, negative consequences have associations too. We might destroy or hide photos of an ex-spouse after an especially painful divorce; looking at them hurts. Because of a strict teacher, Skinner was anxious about misspelling words all his life.[21] Personally, I take the time to cover up notes of things to do on my desk when I'm writing; I'd rather avoid being reminded of all that work awaiting me.

Descending further, we have all done something stupid and suffered waves of public humiliation—often still unpleasant when remembered years later. If the boss at your old summer job was a disappointed witness, you might have found yourself sweating and feeling miserable the next time you saw her. Admitting "I was wrong" tends to accompany unpleasant emotions, which can become classically conditioned as a result. That can be an additional reason why many of us avoid accepting blame even when we know we should be strong and take it.

Knowledge of these associations comes in handy. As a teacher, I avoid red ink when I comment on student papers to try to minimize the negatives associated with it. One of my graduate school friends took advantage of emotional associations in a particularly clever way. When she got tired but still needed to study, she donned formal dress associated with excitement and positive emotions. Better that than resorting to energy drinks.

Emotional associations can drive empathy—and sometimes kick it into overdrive. I've been known to leave the room when someone in a TV drama gets humiliated. Yes, I know the show is not real, but my emotions are. As a kid, I treated some stuffed animals as if they had feelings, and that's rather common. Novelist Jonathan Franzen noted that, as a self-centered child newly awakened to the emotions of others, he began to feel sad for board games he never played. Sometimes he opened them so they wouldn't feel lonely.[22] (This "pathetic fallacy" acquired a name only in the mid-nineteenth century, but doubtless traces back for millennia.)

Properly placed empathy may be our most important human skill. Thus we benefit from a wide range of experience, whether real or imagined. On hearing from her son that reading had made him cry, *Best American Short Stories* editor Katrina Kenison wrote: "So, I thought, now he has been through this rite of passage, the discovery that words on a printed page can give rise to such intense emotion—that a *book*, of all things, can move you right out of your own comfortable little self and into someone else's pain."[23]

Again, the neurophysiology runs in parallel. When husbands were given shocks while their wives watched, the wives' brains registered pain on an fMRI, and their ratings of the painfulness of the experience correlated with their brain responses.[24] Similarly, married couples watching tapes of their own conversations experienced the same sort of physiological effects then that they had had at the time.[25]

There are some offbeat applications of emotional classical conditioning. Andrew Zimmern, host of the Travel Channel's *Bizarre Foods*, eats termites and tarantulas, and people love to watch.[26] Why do thrills of delicious disgust go with a rush to turn on the tube? Perhaps it's like marveling at trapeze artists performing without a safety net (more gripping) or at the brave souls facing danger on CBS's *Survivor*. We can put ourselves in their place and enjoy the adrenaline rush in safety.

An unexpected danger of this emotional conditioning was also mentioned in chapter 6. With so much real-life horror on TV, in our newspapers, and on the Internet, our normal classically conditionable emotional responses can habituate. After repeated presentations, a sudden loud noise will fail to elicit your startle response. Similarly, while pictures of starving children can inspire compassion in a way that statistics can't, too many such images dull the effect. We need to find ways of maintaining compassion regardless, and of learning how to feel it even when looking at those dry statistics. Doing so will require a better understanding of how classical conditioning and consequences work together (see chapter 16).

VALUE, ANTICIPATION, AND LEARNED CONSEQUENCES

Classically conditionable responses interact with consequences in a variety of ways. For instance, when a breeze picks up, your reflexive eyeblink can be overridden by consequences. ("I'll give you $10 if you don't blink." That's manageable unless there's a howling gale.) The withdrawal reflex pulls our hands away from hot stoves. But if your two-year-old trips and falls into the campfire, consequences trump your "Are you nuts?" reflexes, and you will suffer the burns to pull her out.

More frequently, Pavlovian effects tango with consequences by changing their value. Taste aversions offer an unfortunately common example: Eat some spoiled meat and you may shun hamburger joints for quite a while. Young blue jays that had sampled poisonous monarch butterflies retched later just at the sight of one.[27] What aspects of a food will become negatives to be avoided? Research suggests that proteins and dairy products are more likely than carbohydrates to court taste aversion. So, for example, the bacon in a BLT is more likely to be conditioned than the bread.[28]

Fortunately, this effect works positively as well. Your fear of enclosed places like caves might be overcome if your best friend talks you into visiting one together and you have a great time. This change of association is sometimes used in therapy for phobias (see chapter 15). Along with the reduction in fear comes an improvement in consequence value. As a result, the next time you have a chance to tour a cave, you might be less reluctant.

Environmental features, internal cues, and behaviors themselves can all become classically conditioned eliciting signals, like Pavlov's bell—much like the wide range of signals that can be associated with consequences (see chapter 7). Some signals get classically conditioned more easily than others, but there's more flexibility than has sometimes been thought.[29]

Even background context can be classically conditioned, as we saw with drug tolerance. After a severe case of food poisoning, being near the problem restaurant can make you feel queasy. In the same way, if you just got seasick, the mere sight of a boat can bring a touch of discomfort. Extensive animal research verifies the context effect, and it applies broadly to classically conditioned reactions.[30] In one study focusing on an unusual response, a dog repeatedly put in a medium-temperature room before being moved to a hot room showed a decrease in metabolism, simply elicited by the waiting room (no bell needed). Reversing the room order reversed the effect.[31]

Like context, time is another overlooked factor that is invisibly but firmly linked to behavior and consequences (as in chapter 7)—and classical conditioning. If given a mild shock every five minutes, we start tensing up shortly before one is due, even if we are distracted and not consciously aware of the time (and animals tense up in the same way[32]). Mondays feel different than Fridays. A vacation every June brings pleasant spring anticipation.[33]

Get into the habit of *not* enjoying and celebrating what should be big rewards and you risk losing some of these anticipation benefits. That can be a major loss: sometimes the repeated joys of anticipation dwarf the rewards at the end. Many people find the anticipation of a holiday, such as Christmas, to be more enjoyable than the actual day itself. (It's the journey, not the destination.) Emotional anticipation may be the most pleasant consequence of classical conditioning in everyday life.

Building on the previous chapter, signals for consequences often become consequences themselves. Over half a century ago, based on original research by Skinner, a scientist gave water to thirsty rats. Before they drank, a buzzer sounded that accordingly became a signal for reinforcement. Whenever the rats heard it, they could go and drink water. Would the buzzer also become a reward in itself? A lever was now made available, and pressing it sounded the buzzer. No water was provided in this stage of the experiment, the idea being to test the reward value of the buzzer alone. The rats pressed the lever many times, particularly when the buzzer sounded on an intermittent schedule of reinforcement.[34]

Sometimes a signal for a consequence can also be classically conditioned. Both learning processes frequently operate simultaneously. Tiny rat pups smelled a nonfood odor while they were suckling and came to prefer that odor as a result of this pairing.[35] Sometimes one trial is sufficient to create a new reward in this way—a "learned consequence."[36] That is in contrast to something like the taste of sugar, which doesn't have to be learned in order to be rewarding.

For humans, too, learned consequences can waltz with Pavlov. Suppose you find a $100 bill underneath a bush. You start spending more time by that lucky plant; its presence has become rewarding even though it has signaled the presence of other rewards only once (one trial). Where does classical conditioning come in? The day after the find, approaching the bush will likely get your heart rate up a bit and evoke memories that make you smile because of the emotional associations. The signaling—"Free money here!"—is a function of consequences. Again, both types of learning often work together.

We should be especially aware of the power of learned consequences, since most of our own rewards fall into this category. A mentor's approval, career success, new skills like woodworking or fluency in a second language—all these are motivated by consequences that are primarily learned. Money, learned. Jokes, learned. Parking tickets, learned. (They aren't all positives, unfortunately.)[37]

LEARNED AND UNLEARNED

We've already seen that some unlearned consequences can reverse in value if they are paired with, say, food poisoning or shock. Learned consequences might seem easier to alter, but that's not necessarily the case. A learned consequence like money can be paired with all sorts of other rewards (a "generalized reinforcer"), making it especially powerful and hard to modify. Nonetheless, Confederate money became worthless after the Civil War. The social rewards we give each other also tend to be associated with many other rewards: good times together, shared successes, support in getting through hard times. Still, a series of disappointments can turn a loving couple into a messy divorce.

Many everyday consequences combine both the learned and the unlearned. For example, a friend recommends a movie (making it an effective learned consequence), and you make a special trip to see it as a result. You love the clever dialogue about modern electronics, and the dog reminds you of a cherished childhood pet—more learned components. Meanwhile, the sheer sensory stimulation—colors and action and trumpet calls—includes unlearned rewards, much like the kaleidoscopes of chapter 1.

It's not always easy, then, to label a consequence as learned or unlearned. Not only can multiple factors contribute to its value, they can conflict. Alcohol provides a good example, as a learned/unlearned strong positive, strong negative, and everything in between. Sometimes alcohol's value depends primarily on the context in which it is available. Are you among drinking buddies or disapproving relatives? Unexpected contributors include prenatal exposure (at levels well below those that produce fetal alcohol syndrome), which can enhance alcohol's reinforcing value much later, other things being equal.[38] Advertising contributes (positively) to alcohol's consequence value, as does the latest medical report (negatively). If it is associated with enough other negatives, such as alcoholism in the family, that can be enough to turn anyone off.

Spoken words of praise seem less complicated than alcohol: clearly, they're effective mainly through learning. Nonetheless, the tone quality of praise might offer a few unlearned reinforcing features. "Great!" and its kin generalize so well that praise from total strangers can be effective. It's not necessarily so, though: while it may take a long time, "Good job" from a problem boss will eventually lose its value.

"Sticks and stones" can break my bones, and names can hurt nearly as much. In one study, words like "honest" and "sunshine" proved to be effective rewards even presented casually after button presses, so that some participants weren't consciously aware of their status as consequences. Conversely, words with negative connotations, like "lost" and "sick," acted as punishers.[39]

Similarly, college students who thought they were performing a vigilance test watched for a target among a series of signals. Two Pokémon figures were used as initially neutral signals. One was paired unobtrusively with a number of positive-emotion words and images, the other with negative ones. Afterward, participants rated the figures, something they had not anticipated doing. The emotion pairings did influence their ratings positively or negatively in the ways one would expect. Another test showed that the pairings had indeed been unobtrusive; participants remained unaware of the influence of the pairings on their ratings.[40] It's insidious. It should be cause for concern that neutral words paired unobtrusively with positives or negatives can acquire some corresponding emotional value. Like the study participants, we could be influenced without realizing it.[41]

More often, of course, associations are overt, pervasive, and all too effective. When unpleasant images are linked with a human feature, that feature can likewise take on unpleasant emotional overtones. War propaganda across the world is often used to exaggerate features of the enemy, paired with shocking images like torture. Conscious awareness plays a complicated role in these associations,[42] as it does in everyday life, and it's the topic of the next chapter.

Chapter 9

OBSERVING
AND ATTENDING

In a classic series of studies, researchers trotted folks in gorilla suits or attractive ladies through the middle of a videotaped basketball game without warning. Their research participants were focused on counting the number of passes made by one of the teams, but it's hard to miss a gorilla, right? Wrong. *"Observers in our study were consistently surprised when they viewed the display a second time, some even exclaiming, 'I missed that?!'"*

—Daniel Simons and Christopher Chabris,
"Gorillas in Our Midst," *Perception*, 1999

I've been known to ask for help finding something at the grocery when I had been standing right in front of it. (Store associates tend to be understanding.) The consequences of inattention can be a lot more severe, as Simons and Chabris point out: In one flight simulator, landing information was projected on the windshield. Some pilots were so focused on it that they failed to see a plane on the runway right in front of them.

A din of competition constantly clamors for our attention. What determines what wins?

THE MANY ROLES OF ATTENTION

A sudden roar: while you're hiking in a dense forest, a grouse explodes into flight and you instantly jerk your head around, following it. Or you're working on the Internet when a pop-up suddenly chimes and animates, grabbing your gaze and your attention. This unlearned orienting reflex follows Pavlovian principles.

More often, though, what wins our attention follows the consequences that matter to us. We learn to ignore those time-wasting computer pop-ups to focus on what we should be doing (most of the time, anyway).

Despite its insubstantial nature, there is a reason we "pay" attention: the less rewarding the activity, typically the more effortful the attention. No wonder attention can drift toward a more immediately rewarding activity like daydreaming. Toward the end of the workday, this self-control challenge can become particularly daunting.

Neurophysiology findings back up our sensation: as a behavior becomes well learned and requires less attention, for example, fMRI studies show a decrease in blood flow—and therefore less activity—in the relevant brain areas.[1] Along the same lines, people listening to music instead of monotonous background sounds activate their brains. If they focus on different aspects of the music, their brain patterns change.[2] In effect, neuroscientists are starting to be able to track attention in the brain. In related research, neuroscientist Michael Merzenich found that hearing sounds didn't modify the auditory cortex unless there were rewards for paying attention to them (see chapter 4). When attention follows consequences, neurophysiological effects do too. Finally, cognitive scientists have shown that attention helps determine what is remembered, which also changes the brain. As Merzenich concluded, "These choices are left embossed in physical form on our material selves."[3]

It's best to pay attention to what we pay attention to.

We learn to do just that by ignoring what is unimportant. We can focus on just one conversation at a noisy party, for example. Multitasking party-goers can manage attending to two rewarding chats at once. While we all have limits, experts can juggle an impressive number of simultaneous demands. Selective attention doesn't always take a high level of awareness, though: At that noisy party, hearing your own name in a conversation on the other side of the room can be a signal that gets through. (You have learned to respond to it all your life, after all.) Even so, if you're completely immersed in your own discussion, you might miss it. You might even miss a gorilla.

Similarly, when I drive someplace new, I know I have to pay attention to some extent, and I usually retain some memory of the route. When I'm being driven by someone, though, I seldom remember unless consequences make the extra attention worthwhile.

Differences in the level of attention shade into each other gradually, as is the case for other behaviors. At a meeting, we might give no attention, hear but not follow closely, or listen intently. At the other extreme is drifting into a relaxing reverie and suddenly realizing everything's gone blurry. Shifting attention gets our eyes and minds back in focus.

When we focus our attention on each other, another role becomes evident: attention is a powerful consequence in itself. Attention from fellow teens routinely alters hairstyle and dress. Attention for reckless behavior potentially leads to tragedy. Attention from a lover or a beloved mentor can move mountains and transform lives.[4]

Among friends, attention is a generalized reinforcer, associated with all sorts of other good things. Friends frequently reward each other simply by following each other with their eyes. The message doesn't require a smile or gesture or word. The friend doesn't even have to be human. What does your dog do to get your attention? (What do *you* do to get your dog's?)

Anything as powerful as attention has a dark side. Blundering in front of others can make their humiliating attention unbearable. Considerate onlookers look away or pretend they did not hear.

Even seemingly positive attention can have startling consequences. Early researchers discovered that some developmentally disabled youngsters were gouging at their eyes or banging their heads against their desks mainly to get attention. For these kids, when attention was provided for healthier behaviors instead, the self-abuse stopped (see chapter 15).

Similarly, an otherwise-normal four-year-old girl who had been sick continued coughing after she'd recovered. This went on for several months while medical and other explanations were ruled out one by one. Researchers eventually showed that attention from her parents was rewarding the coughing. When they paid attention to other behaviors instead, the cough finally disappeared.[5] (This sort of scientific "functional analysis" deciphers when a consequence is actually influential, answering the "why" questions; see a variety of applications in part 3.)

Attention has yet another role as a surprise reward for ourselves. One joy of life is the state of consciousness known as "flow": becoming so engrossed in doing something that you lose track of time. Photocopying is normally a boring chore, but I once achieved "flow" doing it for a summer job (don't ask me how) and actually worked late without realizing it—not something that normally happened on that job. More often, the involvement is in a more intrinsically rewarding activity, and then even big negatives can fail to receive attention. One naturalist sneaking up on an exciting bird was so focused that he stepped on a large piece of broken glass without noticing—with bare feet.[6]

We learn to selectively attend to our inner events as we do to outer ones. Attention is also fully utilized in "mindfulness," when we pay attention to our attention, objectively detecting our thoughts, emotions, and bodily states. Some religious and philosophical traditions provide extensive training to achieve mindfulness and the consequences it brings.

NOT-SO-SIMPLE OBSERVATIONS

Scientifically examining something intangible like attention can be a real challenge, but researchers have been up to the task. Sometimes they require an "observing" behavior to make attention concrete. For example, when studying perception, they have participants press a button near the screen to make the samples appear for viewing. That way researchers know that participants have paid some attention to them. (Believe it or not, participants have been known to sleep during research sessions.) With a concrete observing behavior, it's easier to see which factors influence attention.

This technique works with animals as well. They also have a capacity for attention, of course, and like people, animals don't always pay attention to what we want them to. (Anyone with a pet knows this.) In chapter 7, rats learned to signal which of two drugs they had received, and then categorized new drugs accordingly—attending to and reporting on their inner signals because they were rewarded for doing so.

In one especially fruitful research line, animals like pigeons could peck either an "observing key" or a "work key." Pecking the observing key produced a signal—a colored light, say. The signal indicated what schedule of reinforcement was in effect on the "work key"—and thus, whether the schedule was richly rewarding or whether the birds might have to peck a long time to get anything. The signal might also indicate that no rewards were possible. In no case did observing the signal actually change the schedule on the work key. Why peck the signal key at all? As in real life, the schedules of reinforcement on the work key might switch around unpredictably, so pecking the signal could save wasted effort. Why peck the work key unless it might pay off?

The big surprise was that, during periods when no rewards were available, pigeons wouldn't peck the observing key much, even though its signals were still providing useful information.[7] After all, observing was the best way to learn whether pecking the work key was worthwhile. And only by observing during periods when the signal was *negative* (no rewards available, say, or a miserably lean schedule) could they pick up on when the signal changed to *positive* (a rich schedule).

While failing to observe may be inefficient, the fact remains that bad news is no fun, and enough of it can make the signal a learned punisher (potentially bringing classically conditioned emotions). Rats, monkeys, and fish also fail to observe under these circumstances.[8] Observing often fails to occur even when the choice is between two rewarding schedules, rather than between something versus nothing. When one schedule is substantially richer than the other, only that one supports much in the way of observing.[9] (Everything is relative.)

We could hope that we humans would be wiser and more willing to take the potential disappointments of observing when it's in our own best interest. Sometimes we do. But the overall message is clear: sometimes we would rather bury our heads in the sand than risk suffering bad news.

This "ostrich effect" is evident in human laboratory research of the sort just discussed[10] as well as in research scenarios closer to real life. In one classic study, participants listened to a series of five-minute messages that were sufficiently masked by static as to be unintelligible. Pressing a button five times cleared the static for three seconds. The participants knew the order of the topics. When would they choose to button press and listen in? The highest observing rate was for smokers listening to a message that smoking was not linked to lung cancer. Their observing rates were 35 percent lower for a more accurate but discouraging message about smoking's link to cancer. Nonsmokers pressed at an intermediate rate for both of these messages.[11]

Accordingly, behavioral economists found that Swedish investors checked their pension portfolios substantially more in rising stock markets than in declining ones.[12] The same thing was true for investors in the large fund group Vanguard. In a particularly striking example, polar explorer Admiral Richard Byrd, alone in the Antarctic winter for months in 1934, did not want to hear potentially bad news about the stock market over his radio link to base.[13] He couldn't do anything about it, but of course he also missed out on the chance of good news during that trying time.

In their book *Mistakes Were Made (but Not by Me)*, Carol Tavris and Elliot Aronson noted how extreme the ostrich effect can be: across the United States, many prosecutors did not want to know if they had helped convict the wrong individual, and resisted DNA investigations that might have cleared the innocent. The prosecutors' jobs and self-images were both at risk; it's a big negative to discover you were wrong when so much was at stake. As Tavris and Aronson point out, a number of states do not compensate the wrongfully convicted, even after years in jail. Some states do not even expunge these convictions from the official record, making it that much harder for the wrongfully convicted to try to get on with their shattered lives.[14]

Reassuringly, when the consequences are powerful enough, we often do risk pain to check for potentially bad news: we bravely get tested for cancer, then, bracing ourselves, we call the doctor's office for the results. Even with our own lives at stake, though, sufficiently delayed and uncertain consequences can fail to offset our immediate fears. Hundreds of thousands of Americans are estimated to be infected with HIV without knowing and without wanting to know, despite the existence of an inexpensive test.[15] Since they can carry the disease for years without developing symptoms, they can be infecting others in the meantime. Like the cigarette smokers, they are rolling the dice, not wanting to observe and learn the truth.

Historians still debate whether Albert Speer, Hitler's minister of war production, knew about Hitler's "final solution" for the Jews. Whatever the case, Speer said later that "If I didn't see it, then it was because I didn't want to see it," and he alone at the Nuremburg trials accepted responsibility for the war crimes of the Nazi regime.[16]

BENEATH THE RADAR:
CONSEQUENCES WITHOUT AWARENESS

Knowing our tendency not to want to know can help us fight it. But even when we might want to know, in some cases we can be influenced by consequences without our awareness. We have seen how students innocently chatting with researchers were unaware that they were being tracked onto family or nonfamily topics by rewarding grunts (see chapter 7). Other researchers using the same basic method have confirmed this finding, as have a variety of rather different research lines.

Norwegian psychologist Frode Svartdal, for example, had his participants thinking they had to reproduce the number of clicks they had heard (not easy). Actually, they were rewarded only for responding within a certain time window, either fast or slow. Their response times followed the rewards, but they were unaware of the relationship, claiming that it was their accurate counting that did the trick.[17] They were influenced by the consequences without realizing it.

In a follow-up, Svartdal distracted his participants while they were trying to perform an easy or difficult task—but reinforcement depended simply on the force of their button-pressing responses and not on the task as they had been told. Only for the difficult task did the force of their responses track the rewards.[18] Attention to inaccurate rules about getting consequences can thus actually interfere with our sensitivity to those consequences (as we will see in the next chapter).

In yet a different approach, two researchers hooked up volunteers to electromyography equipment that could record tiny muscle twitches in several areas of the body. The participants were simply asked to relax throughout the session and told that they would earn money (exactly how was not specified). After the participants had settled down, the researchers rewarded otherwise-imperceptible thumb twitches of a particular size range. Each nickel earned was displayed. The muscle twitches duly occurred more often during this phase and decreased when the rewards stopped. Meanwhile, muscle twitches in other size ranges either stayed constant or declined. Restoration of the reward requirement soon restored the desired thumb twitches. Afterward, none of the participants could say what they had done to earn the money, only that they were annoyed when the rewards stopped coming.[19]

IT'S AUTOMATIC

We may be influenced by consequences without being aware, but can we *behave* without realizing? To a surprising degree, the answer is also yes. Riding a bicycle requires intense attention at first and plenty of obvious consequences (sometimes rather hard ones). Later, it becomes automatic—so well learned that little or no attention is required. The same is true of many routine activities, like toothbrushing. In contrast, behaviors that do require attention, like contributing to a serious conversation, are "controlled." In between are intermediate levels.

Some unaware behavior falls into a separate category altogether. A sudden unthinking snatch of your hand out of a flame is the withdrawal reflex of chapter 8, not learned, and mediated by the spinal cord rather than the brain. Stopping at a red light seems reflexive too, but of course it is not. Like riding a bike, it's learned through consequences. Experienced drivers coming to a light that is just turning red can brake without an interruption in their attention elsewhere. New drivers have to learn this automatic braking reaction, and brake pedals in many vehicles are standardized to help ensure it works even when you're driving your brother Ernie's car. It is so well learned that if you are in the front passenger seat in an emergency, your right foot might shoot out to slow down the car even though there's no brake pedal on your side.

For most of us, driving becomes so automatic that we switch to auto-pilot and start daydreaming even while cruising the freeway and passing semis. It's a good thing that we really aren't lost to the world like it seems. Consequences—and their signals—still affect these behaviors, and we shift attention very quickly when a police car appears or a reckless driver cuts us off.

B. F. Skinner suggested that reading can be automatic,[20] and I know I read billboards along the freeway in this manner. Even when I become aware that I'm doing it, it can take a conscious effort to look at something else instead. In this case, not only is reading a well-learned behavior that I do frequently, it's being rewarded on a variable work-based schedule of reinforcement: occasionally a billboard is funny or attractive or informative. It's like gambling or reading the newspaper—it can be hard to stop because the next toss might win, the next article might be great.

Skinner also noted that extemporizing on the piano was most successful when he found a happy medium between controlled and automatic behaviors. Some awareness was required, but too much conscious planning for chord changes slowed things down.[21] In the same way, improvisational theater or jazz is a blend of the consciously controlled and the well-learned automatic. When it works, the effect is brilliant.

We learn to take advantage of the benefits of automaticity. Martial-arts experts automatically fall so as to avoid hurting themselves—something that takes a lot of practice. Kayakers, like me, practice bracing with a paddle in case a rogue wave threatens. In strong winds, my automatic bracing stroke—one so well learned that I don't have to pay attention to it—is almost as important as my life jacket.

Some behaviors can become automatic without practice. Standing in a banking line that's going nowhere, reviewing plans for that evening, we automatically shift our weight whenever the discomfort of standing builds up, completely unaware that we're doing it.

This attentionless ease has its dangers. Automatic habits are still modifiable, but changing them is harder than if they weren't so well learned. Conscious effort can be required just to notice that they are happening, as in the banking-line adjustments. When I'm driving, I sometimes make a wrong turn because it's where I normally turn, and my attention is elsewhere. I didn't plan to go on autopilot, but the normal route was too well learned. It can be annoying. And embarrassing: newscaster Ed Meyer of Washington, DC, used to speak on a morning show, then leave on his assignment for the day. Once, when the news director gave him his assignment early, he automatically headed out, only to realize, stuck in traffic, that he hadn't done his broadcast yet.[22] Oops.

More seriously, ex-smokers can still find themselves reaching for a pack without thinking, even months after quitting. Past consequences for years have made the behavior automatic, particularly in situations that have signaled it. Two thousand years ago, the Greek philosopher Epictetus noted:

To make anything a habit, do it.

To not make anything a habit, do not do it.

To unmake a habit, do something else in place of it.[23]

This is still good advice (more on how to carry it out in part 3).

Offhand, it may seem that most of what we do is controlled, simply because that is what we pay attention to (and therefore remember). Instead, social psychologists researching our daily lives have found that automatic behaviors such as habits predominate.[24] It's best to be aware of them.

OBSERVING OTHERS

Being aware of others and the useful things they do can save a lot of painful individual learning. Why reinvent the wheel? This form of learning from consequences expands the ability to learn *about* consequences. At its basis, it's a matter of learning what to pay attention to.

Flies gather on a carcass not because they are observing each other but because they are all following the scent. In contrast, vultures gather on that same dead animal because they have watched each other, circling in the sky. They've learned that when a neighbor heads down, rewards are probably waiting on the ground.[25]

It's surprisingly easy to learn through observing others. In chapter 2, wasps learned to seek caterpillars hidden in leaf shelters only after they had eaten a caterpillar in a partially demolished shelter. They got the signal then. In a similar way, when trained pigeons pierced a paper cover to get at grain underneath, birds who watched were able to pierce the paper themselves and eat. Birds who watched models eating without having to pierce the paper were mystified when presented with a covered container.[26] It's a straightforward extension of ordinary signal learning. Indeed, sometimes the actions of models simply make a feature more obvious. Young house mice follow their mothers in the nether regions of your kitchen and help eat what she finds—and that helps them learn about what's good to eat and where to find it.[27] Similarly, when some monkeys are polishing off a new kind of fruit with relish, others have learned that it may be worth trying—they've been rewarded for doing so in the past. It's like copying a friend who dips into an exotic dish at a Chinese buffet.

If spending time near a parent or another model is indeed rewarded, all these things happen naturally, at least if the relevant consequences are powerful enough. When Japanese quail were well fed, for example, they did not observe a demonstrator bird getting food in another chamber (by either stepping on a pedal or pecking it). Later, when they were themselves tested when they were hungry, they showed no learning. However, when the observers were hungry, they actually did observe the demonstrator and the food, and later benefited from the experience.[28] Observing doesn't just happen: as we saw earlier in the chapter, it is itself a behavior affected by consequences.

Learning through observing is common in the wild. Young oyster-catchers, a type of shorebird, watch their parents cut the locking muscle of an oyster with their beaks, or, in the direct approach, smash the tough shells. Youngsters can stick around for months, watching, trying, and frequently failing. Other oystercatcher parents specialize in easier-to-get worms, and their young pick up the techniques more quickly and become independent sooner. Oystercatchers aren't stuck for life with what they learn from their parents, though; they can learn on their own or by observing unrelated birds. (Some even ditch the honest way of life and find success as pirates, stealing food from other birds.)[29]

Cheetahs take learning through observing a step further. Cubs watch their mother catching prey and share in the rewards. When they're older, Mom captures a young gazelle but doesn't kill it, instead releasing it near them. If they cannot catch it yet, she demonstrates how. Repeated failure and occasional success in these practice hunts make the cubs increasingly independent, in a shaping progression that involves instinctive as well as consequence-based behaviors.[30]

Observing can expand to establish a culture. On an island, a colony of Japanese macaque monkeys was maintained so zoologists could study them. A one-year-old female, Imo, famously began washing sand off the sweet potatoes that got dumped on the beach for them. Other youngsters who played with Imo began doing the same, then her mother, then their mothers, and finally the adult males. Several years later, Imo discovered how to separate wheat from sand by tossing a handful into fresh water. The sand sank, while the wheat could be picked off. Again, this practice became a tradition, and both traditions have now been maintained for decades in this troop.[31]

Some of these cultural practices include tool use. A unique foraging tradition among some chimpanzees traces back 400 years—opening nuts by using a stone as a hammer.[32] More recently, both female and male chimps have created spears to successfully hunt bushbabies.[33] Orangutan cultures offer a range of varied tool uses, including leaves as gloves for handling spiny fruits.[34] In one bottlenose dolphin pod, the females (and occasionally males) wear sponges over their snouts when they are feeding by probing, perhaps to protect themselves. The practice appears to be passed along from mother to daughter via learning through observing.[35]

Even birds have cultures based partly on learning through observation. Another well-known example comes from the famous New Caledonian crows that make as well as use tools and pass the skill along the generations (helped, it appears, by an instinctive component). Mated pairs often stay together for life, and the young birds remain with their parents for about a year, often longer. The families frequently forage together, with young birds observing their parents making tools and using them to work grubs and other goodies out of crevices. The young get to eat what their parents secured using tools, and they get to try using discarded tools. Still, even after six months of training, they are often unable to make a tool successfully. Researchers observing a number of family groups on their native island concluded that it takes a combination of observation plus a long series of individual successes and failures. By one year of age, the young can use tools effectively, although they're still a long way from matching the expertise of their parents.[36]

THE ULTIMATE IN OBSERVING: IMITATION

While many of these observing-based behaviors look like imitation, most would not actually qualify. Those vultures descending to the ground aren't truly imitating each other, for example: the similarity between the flying descents doesn't occur for the sake of matching, it's just incidental. Even in the case of the Japanese monkeys, it's hard to say to what extent true imitation occurred, rather than other forms of learning through observing. Only careful experiments can rule out the alternative explanations.

Such experiments have shown that, while people are the paragons of true imitation, we are not alone in possessing this formidably useful ability. We have known for many years that chimpanzees are among the species that can learn to imitate, for example. In a recent study, two chimps were at first rewarded with praise and food for imitating gestures like slapping their hands on the floor, behaviors they already did occasionally. No food was then given for tests of new gestures—although the researchers had to reinstate it periodically after the chimps realized no rewards were coming and staged a strike. Independent observers recorded what the chimps did without seeing the model. The chimps didn't imitate everything by any means, but it was clear that they were able to copy at least some aspects of many of the movements.[37]

Not surprisingly, kids who are rewarded for imitating catch on quickly. Psychologist Claire Poulson and her team worked in depth with three twelve- to fourteen-month-old infants and their mothers. At first, the moms just modeled actions like clapping, or made a simple sound like "ooh." Babies were asked to "do this." The researchers recorded whether the babies imitated, but there were no consequences for doing so—and very little imitation occurred. Then the moms provided praise and a small treat, and imitation steadily increased until it was reliable. Finally the critical tests: new movements and sounds, and no praise, alternating with standard rewarded tests. Would the babies imitate? Indeed they did. They had learned.[38]

Like many seniors, B. F. Skinner delighted in these sorts of games with his five-month-old granddaughter. He noted that when he held up his left hand, fingers outspread, she matched his gesture with her right hand—"mirroring" rather than truly imitating.[39] Researchers explored imitation and mirroring in two- and three-year-olds, rewarding correct imitations of training gestures on an intermittent schedule, so not all correct imitations were reinforced. That meant test trials of new gestures were more unobtrusive. While three-year-olds did better than two-year-olds, both groups frequently made the error of mirroring rather than imitating gestures made with one hand. The children also tended to use their dominant hands regardless of the proper choice, and they were better at gestures that are commonly rewarded in preschools and kiddie TV shows, such as touching their ears.[40]

There is intriguing evidence that "mirror neurons" can activate when we copy an action or see others copying. However, the degree of activation reflects experience. For example, ballet dancers showed more mirror-neuron activity watching ballet dancing than martial arts, while martial-arts experts showed the opposite pattern. Other people had little mirror-neuron activity to either activity.[41] Once again, our brains get shaped by our experiences—including consequences—and of course, many other elements of our neural systems participate in imitation, as they do in any behavior. The full story of mirror neurons remains to be determined.

To close this chapter, my favorite imitation story: "Clicker training" expert Karen Pryor brought positive reinforcement to a zoo with a bored female orangutan. One of the keepers used food to train the orangutan to imitate. It became so much fun that the animal spontaneously started imitation games with children passing by, to great acclaim.[42] The consequences for the orangutan no longer included food but evidently did include that ubiquitous reward, attention.

Chapter 10

THINKING
AND COMMUNICATING

> *"[A] domesticated chimp, Viki Hayes, was given two piles of
> pictures, one of humans, the other of nonhumans, and then handed
> a stack of additional pictures and invited to categorize. Her
> performance was perfect, with one small exception: She placed the
> picture of herself among the humans."*
>
> —Carl Sagan and Ann Druyan,
> *Shadows of Forgotten Ancestors,* 1992

It's hard not to feel touched by Viki's choice. All the same, it makes perfect sense. She spent most of her time with people; why shouldn't she be in the same category?

Consequences are a major basis for categorizing. Two pigeons successfully learned to peck a red triangle, not a green circle. Follow-up tests showed that one bird learned on the basis of shape, the other based on color.[1] Fine-tuning the consequences can create standard categories and not leave things to chance or idiosyncratic histories. Told that odd-looking objects were toys, preschoolers in one study grouped them based on shape rather than the other obvious choice, color. They had learned that shape distinguished toys like cars, while color was usually an incidental feature.[2] Using shape let them generalize correctly to new examples of toy cars (pleasing their parents no end).

It's a learning process: when toddlers overgeneralize the "daddy" category to grandpa, or "doggie" to the family cat, parents are there again to help. Undergeneralizing happens too. The nineteenth-century "wild boy of Aveyron," brought in from abandonment, applied "book" only to the actual book with which he had been taught. Once his teacher praised him for applying the word to other books, he promptly overgeneralized instead, indicating that a newspaper and a few sheets of paper were "books."[3]

If only good manners would overgeneralize.

CATEGORIES LARGE AND SMALL

How do we draw the line between over- and undergeneralizing? We stop at red lights of different shades, but not if they're green. We read different capital A's, but puzzling out a strange-looking one in a fancy restaurant makes us aware of how effortless generalizing usually is.

The basis for a category can be learned or unlearned, artificial or natural, and tightly defined or fuzzy. Mimics of poisonous animals take advantage of the difficulties of fine-tuning. Because animals that eat monarch butterflies get sick, tasty but similar-looking viceroy butterflies are left alone (see chapter 2). Similar artificial colors and patterns have the same effect.

The results of such artificial categorizing can be surprising. In chapter 1, pigeons that had learned to distinguish eighteenth- and twentieth-century classical music then categorized novel excerpts like we do. Similarly, pigeons learned Picasso versus Monet, again generalizing correctly to new examples.[4] (Upside-down pictures threw off categorizations of Monet but not of Picasso.) These stodgy-looking birds can handle categories like cars, trees, fish (even underwater), and the presence or absence of a particular person in photos, however large or small.[5]

A bigger challenge: Divide forty slides of trees randomly into two sets of twenty. Reward yourself for choosing slides in only one of these arbitrary "categories." Now reverse sets. Like people, pigeons come to treat members of these differently rewarded sets as if they were equivalent, and they learn to adjust quickly after a reversal—a skill long considered beyond them.[6] At one point, learning abstractions like "choose what's different" was also thought to be an elite skill. These days, pigeons are one of many successful species able to do so.[7] Even honeybees can handle the abstraction of "same" versus "different" when they are rewarded for learning it.[8]

In nature, animals choose louder songs or bigger, redder berries, and they readily learn such abstract relations in lab tests ("choose bigger over smaller").[9] A tougher challenge: Dawn is taller than Cheryl, Cheryl is taller than Nancy, so Dawn must be taller than Nancy. Is the recognition of this relationship limited to masters of logic? That four-year-olds could do it was the initial surprise (masters of logic they are not). After that, monkeys, crows, rats, pigeons (of course), and even fish were successful.[10]

Such flexibility foreshadowed the developments that led to sophisticated language.

SIMPLE COMMUNICATION

Animals categorize, master abstract relations, and learn from observing (see chapter 9). Simple communication is a walk in the park by comparison. The rewards for exchanging signals are so large that systems have developed using every conceivable means, including animal versions of sonar and radar.

These systems can be surprisingly complex even in invertebrates. Patent-leather beetles are devoted moms and dads, communicating with their young using different sounds with different meanings.[11] Pretty impressive for an instinctive system in a bug.

To realize the full potential of communication, though, consequences supported more flexibility than instincts ever could. But does this form of learning offer enough flexibility for a horse to do arithmetic? In the early 1900s, the horse Clever Hans apparently could solve math problems (for carrots and apples), answering by moving his hoof the required number of times. Incorrect answers weren't rewarded, so the horse was motivated to answer correctly. Scientists were skeptical, but the horse did well even when other people gave Clever Hans the problems and his trainer was out of sight.

The solution? Upon investigation, psychologist Oskar Pfungst found that Clever Hans got the right answer only when the questioner knew it: questioners were unsuspectingly giving little cues such as head tilts when the horse reached the correct number, and the horse was clever enough to notice. Pfungst tried playing the part of the horse himself, and successfully read subtle cues from about 90 percent of his questioners. Even questioners who knew about the Clever Hans effect still gave these signals involuntarily, communicating quite effectively despite their conscious intentions not to.[12]

Whether aware or unaware, we all give and receive nonverbal cues constantly: smiles and other expressions, gestures, eye gaze, posture. Growing up in Chicago, I learned to walk confidently to signal that I would be a poor target for crime. Many such communication "rules" don't have to be explicitly taught because we learn them through natural consequences. For example, attention is a common reinforcer. When listeners look away, their lack of attention has the opposite effect on speakers. (Well, most speakers.)

Animals are naturally attuned to nonverbal signals, and our pets can learn to read us so well it's spooky. That's how your dog may know you're about to go for a walk before you do.

Our pets can even speak after a fashion, barking or meowing on cue for a reward, an ability shared with many other mammals and birds. Also straightforward is sounding off in different ways in response to different signals for consequences (rather like teens swearing with their friends but not with their parents).[13] Modifying a natural sound is more challenging, but again manageable by a surprising number of species—budgies, for example. Shaped by food rewards, three of these little parakeets learned to give two new, unnatural calls to two different cues.[14] Given their ability to mimic human words, that is not all that surprising. More surprising was a well-documented pet seal that imitated his human parents.[15] (Also recall the chapter 2 example of finches that modify their songs to avoid a negative.)

Most primates lack such vocal agility. Gibbons sing duets, but consequences don't seem to play much of a role for them.[16] However, field experiments showed vervet monkeys in the wild used different sounds to mean "leopard," "snake," "eagle," "baboon," "other dangerous mammal," "unknown human," "dominant monkey," "subordinate," and "rival." While still largely instinctive, this impressive symbolic communication system appears to involve some degree of learning from consequences. For example, baby vervets developed the correct responses to particular alarm calls only after several months, and they were shown to learn from observing adults.[17]

THE UNDERSTANDING ANIMAL: SIMPLE LANGUAGE

In the wild, animals do fine with their existing communication systems. Given training that builds on their abstract categorizing skills, though, the possibilities for more sophisticated language expand considerably. And this is true for more than just apes and dolphins.

Closer to home, dogs and people have been communicating for millennia using gestures, words, wags, and barks. Some dog owners have to spell out words in conversation, because their pets have learned what "walk" and "park" can mean—just like parents of toddlers spelling "c-a-n-d-y." Just how far can dogs go?

Retired college professor John Pilley spent three years finding out, intensively training his enthusiastic female border collie, Chaser, using petting, attention, and play as rewards (seldom food). She learned made-up names for more than 1,000 different objects. (Yes, you read that right, 1,000 with three zeroes. The vocabularies of the world's simplest human languages aren't much larger than that.) This impressive animal learned a number of different verbs as well.

The big test: Given random combinations of three verbs—"paw," "nose," and "take"—and three objects, Chaser was able to perform correct actions with correct objects when asked. She learned names for larger function and shape categories for some of the objects (such as "toy") even when they overlapped (both "ball" and "toy"). She was also able to learn the name for a new object just because the other objects present were familiar.[18] Whew. To eliminate the possibility of Clever Hans–type visual cues during testing, her teacher gave the commands when he was out of sight.

In comparison to the vervet monkeys, Chaser demonstrated "referential language," where words refer to objects independently of associated actions—a skill once thought to be restricted to people. Well-controlled studies by psychologist Louis Herman have demonstrated referential language in dolphins as well. Using a gesture language of forty words that included objects, locations, and syntax—word order—many different actions and orders could be tested. A dolphin called Akeakamai passed with flying colors, handling sentences of up to five words, and a larger range of combinations than Chaser.[19]

While both Chaser and Akeakamai could be said to understand language to some degree, producing it is another matter. Most impressive, then, are the achievements of primatologist Sue Savage-Rumbaugh's Kanzi, a bonobo that made stone tools (learned by observing an anthropologist) and enjoyed playing the old video game Pac-Man. As an infant, Kanzi learned the basics of language through observing his mother being taught. He achieved a vocabulary of over 200 words, and in addition to understanding spoken words or symbols, he could select symbols to "talk." He handled word order just fine. He even invented meaningful sounds of his own to go with some of the symbols.[20] Despite these accomplishments, Kanzi and other apes in language programs generally put together only two- or three-word sentences.

HUMAN LANGUAGE AND ITS CONSEQUENCES

Our large brains helped us do better, and so did a change in the position of the hominid larynx: while it made us liable to choke (unlike other primates), it brought extra flexibility to our speech.[21]

Speech has significant advantages over the gestures that some ape cultures use: for example, you can be working with your hands while you speak, and you don't have to be in line of sight to hear. B. F. Skinner thought the way our speech came to be flexibly modified and influenced by consequences was another critical change.[22] These two developments helped open the floodgates: even the simplest of the world's thousands of languages has vocabularies of over 1,000 words and complex functions galore—functions with consequences.

Language consequences are everywhere. We ask outright for some. ("Can I have a chocolate shake?" "Do you know how I get back to the freeway from here?") We chat with friends, persuade the boss, practice a speech—all for consequences. Conversations are like tennis volleys, and the socially skilled rise to giddy heights: Not saying everything they know, so that someone else can share the rewards. Deliberately misunderstanding something to avoid embarrassment for another. Keeping track of what "he knew she knew he knew." Ensuring no one dominates the discussion, instead keeping the rewards coming for everyone so everyone stays engaged. How do we achieve these skills?

Right from the start, natural consequences support and fine-tune requests. As a toddler, ask for a "tookie," and you've got a shot. Try "huh," and you're out of luck. The same toddler, when asked to name the treat, might be rewarded by receiving praise instead of the cookie. Because the function of the same word is different (commands/requests versus names/descriptors), so are the consequences.

Of these two common functions, requests usually benefit the speaker while descriptors more often assist the listener. But there are mutual consequences for all concerned. Comment that it has started to rain and be rewarded by "Thanks, I'll take an umbrella when I go out," or at least an interested look out the window (attention again as reward).

SAME WORD, DIFFERENT CONSEQUENCE

Because the functions of requests and names are different, Skinner made an unusual prediction in the 1950s.[23] If a hungry toddler can request a cookie, it would seem she could name it when asked—regardless of whether she was hungry. Similarly, naming ought to transfer back to requests ("What is this?" versus "What do you want?"). Skinner suggested that, on the contrary, this transfer might not be automatic, because the consequences, motivations, and associated signals (like the question) were different. At the time, however, no experiments had been done.

In the first one, primatologist Sue Savage-Rumbaugh and colleagues taught chimps requests for different foods. Would the words then function as descriptors? That is, would the apes be able simply to name the foods when asked? The answer was no, not without separate training.[24] That's chimps, though; what about people? Suggestively, the nineteenth-century "wild boy of Aveyron" had trouble with the opposite direction: he had learned names without being able to use those names as requests.[25] Again, the words were the same but the consequences were different. But that was just an observation, not an experiment.

However, the first human experiment also showed that ordinary pre-schoolers needed extra training to make the transfer. They learned "on the left" or "on the right" as either requests or descriptors, as in "Where do you want me to put the flower?" versus "Where is the flower?" Although training was successful for either function on its own, "on the left" and "on the right" failed to transfer automatically in either direction: kids who learned them as requests couldn't handle the same phrases as descriptors; kids who learned them as descriptors could not give requests.[26]

In a follow-up using concrete objects instead of "left" and "right," kids who learned requests *could* use the words as descriptive labels, but the opposite direction did not work.[27] Subsequent research has validated Skinner's prediction many times in normally developing children and in developmentally disabled children and adults—and such research has helped lead to better ways to teach this important transfer skill.[28]

It's really a form of generalization: the same word for the same object, but for different consequences, to different signals. Our hungry toddler learns that "cookie" works for the same object as a name ("What is this?" "Cookie." "That's right!") and as a request ("What do you want to eat?" "Cookie." Receives cookie, munching sounds.)

In fact, just learning how requests work is a form of generalization, too. Requests are "frames"—repeated language structures like "can I have"—and once a frame is learned, it is easier to generalize to new requests. Because of frames, we can make some sense of nonsense statements like "the tok was plithing." We generalize words, then we learn to generalize frames.

Simpler frames can be just parts of words: forms like "-ed" for past tense and "apostrophe *s*" for possessives. (Frames can get really complex, too, as in analogies, metaphors, and mathematics.) A child learns "the boy's shirt," "the boy's shoe," and generalizes to "the boy's coat." But then, if "-ed" works for "walked" and "talked," what's wrong with "goed" and "runned?" Research shows that generalization commonly produces these forms until kids learn all the exceptions. In some kids, the correct form for an irregular verb actually appears first—it's been heard, after all, and rewarded—but then the general rule is learned and irregulars get placed into it ("runned"), and only later do the irregulars get sorted out properly for good. Correct and incorrect forms can alternate for a while during this process, as kids learn what works and what doesn't, experiencing the consequences.[29]

As researcher Michael Tomasello showed, across different languages, young children's grammar is usually poor, developing slowly and "piecemeal" over a lengthy period.[30] It's not just kids who struggle, of course. Adults learning second languages can relate. But for sufficiently strong consequences, some adults manage to achieve native-like fluency in a second language. Indeed, some linguists have found that adults can learn a new language faster than children can. It helps when the basic language structures are similar, providing more basis for generalization.[31]

BABBLE ON

When I learned Spanish, I lamented the travesty I made of rolling my *r*'s. Baby babbling includes the basic sounds needed for the world's languages, but my rolled *r*'s didn't make the cut. Similarly, Karen and Bob's infant makes different sounds than does Xue and Wei's baby on the other side of the planet. Neither baby can speak a word yet, so what's going on? Scientists have shown that the consequences Mom and Dad offer—such as smiles and praise—shape infant babbling, selecting the building blocks of each language. Demonstrably, this happens even when parents aren't aware of it.

Right from the start, babbling is affected by consequences. In one early study, thirty infants as young as two months babbled away, and researchers rewarded either vowels or consonants with smiles, approving sounds, and stroking. An independent observer categorized the sounds. The babies responded by increases in either their vowels or their consonants, tracking the rewards.[32]

In more recent research, moms and infants played together. When some infants vocalized, moms immediately smiled, moved closer, and touched their babies. Other mothers provided the same "consequences" at the same times but independently of what their babies were doing (so not true consequences). Only the babies who had been truly rewarded for vocalizing significantly increased their babbling.[33] Other scientists have shown that these sorts of interactions occur naturally: in one study, mothers interacting freely with their babies immediately rewarded babbling 70 percent of the time.[34] (Thanks to other research lines, we can assume that their smiles and sounds were probably effective rewards.)

Children get very heavily rewarded when their babbling turns into "Ma-ma" or "Da-da," and most kids pick up those terms quickly. Imitation and generalization help a great deal. In chapter 9, infants readily imitated gestures when imitation was rewarded. The same goes for sound imitation. Three preschoolers who were rewarded for imitating English words, for example, imitated Russian ones, too, even though they were never rewarded for those. When rewards were provided instead for any nonvocal behavior, all imitation dropped. Reinstatement of the rewards picked it up again.[35]

Simply listening to language is clearly *not* enough to pick it up. Interactions—and the consequences that necessarily go with them—are critical. For example, a hearing child raised by deaf parents spent most of his time at home. The TV was kept on for him on the theory that this exposure to spoken language would suffice for his language development. By age three he had readily learned the sign language that his parents used but could not understand or speak English.[36] There had been no consequences for learning English, but there had been plenty for learning sign language. Similarly, people living in foreign countries can be immersed in a different language, but without the motivation to interact and learn, they may pick up only a few words.

LANGUAGE LEARNING IN REAL LIFE

While lab research and case studies are important, large samples of real-life language learning have been especially illuminating. Kids and parents two-step their way through a language dance, and a lot of what two- to three-year-olds say has been shown to receive feedback. Young children hear and speak thousands of words per day. What are they hearing and saying?

My acquaintance Ernst Moerk analyzed many hours of recordings of a toddler, "Eve," chatting with her mom at home. Eve heard each major sentence type about 100,000 times per month, and her mother corrected or expanded her grammar many times per hour. Moerk documented forty different categories of mom's natural teaching techniques, including repeating, asking Eve to repeat, elaborating on what had been said, labeling, simple direct (presumed) reinforcement, and varying the grammatical frame while repeating the content. Parents characteristically also use "motherese" or "fatherese" when talking to infants, simplifying the mix of signals, consequences, and frames.[37]

As a sample, Mom asked, "What is the child doing?" Eve answered, "Running," and Mom agreed with an approving tone, "She is running." Moerk's analysis of the likelihood of patterns like this—Mom's question, Eve's response, mother's consequence—showed the actual occurrence to be up to 100 times what would be expected from chance. In a simpler pattern, Eve's imitation of a word was reinforced up to 50 times more often than expected from chance.[38] Indeed, attention itself (not recorded) is a reward for language as it is for other behavior, and most kids go through a phase where they're talking a blue streak. Eventually, consequences teach them that there are times to talk and times to be quiet.

The most ambitious research project on language acquisition was summarized in psychologists Betty Hart and Todd Risley's book *Meaningful Differences*, which referred to the large skill differences in kids entering preschool programs. These differences made it extremely difficult for the disadvantaged children to catch up. What was happening *before* preschool?

Observers followed children in forty-two homes—a cross-section of America—from nine months of age until they became three years old, recording all interactions for one hour per month. As Moerk had found, parents said, "I don't understand," found gentle ways to correct mistakes, rewarded proper etiquette, and elaborated on kids' comments (for example, in the presence of a car, child: "Car." Parent: "Yes, a green car.") Speaking also brought many natural consequences like answers to questions and (as always) attention.[39]

Hart and Risley took care to point out that all the parents were doing their best for their children. But some were able to provide highly enriched environments, and the effects were clear: based on the recordings, professional parents averaged about 11 million words per year to their kids, while parents on assistance spoke only 3 million words per year to their kids. (Working-class children spanned that entire spectrum.) Children who interacted less had fewer opportunities to learn—and in fact learned less. At age three, during the recorded hours, the professional parents' children were using larger vocabularies than the parents on assistance were. (Note that these latter parents actually had substantially larger vocabularies, but they were not using them with their kids to the extent that the professional parents were.) The more words you can speak, the easier it is to learn to read those words. Differences build.

Another of the critical factors was the positives-to-negatives ratio, closely linked with consequences. In the most talkative families, ratios were high, ranging up to six positives for every negative. In the less-talkative families, the ratio was more like 2:1 *in the opposite direction*: two negatives for every positive, which is discouraging for the kids. Over their first four years, that difference amounts to hundreds of thousands more encouraging words for the kids in the enriched environments.

Factors that made no difference in later language skills included a child's sex, race, birth order, and family size. Because some working-class parents talked a lot and some wealthier parents did not, socioeconomic status was actually a poor predictor as well. So what did predict success? The critical factors did.[40] A good start gave kids a big advantage that lasted years later.

STRICTLY PRIVATE

One early milestone is the switch from reading out loud to silent reading. Some kids vault through it, while others make a more gradual transition from speaking to sounding out just the more difficult words, to silent reading with continued movements of the speech muscles, to fully silent reading. Eventually, when a seven-year-old solves an addition problem out loud—for consequences—and then does a similar problem "in her head," most of the same things are happening in her brain (see chapters 4, 8, 9). As Skinner said, "The skin is not that important as a boundary."[41]

Nonetheless, it does present some special challenges. For example, how do we learn to talk about what we feel inside? No one else can see it and name it, after all.

Our language communities use all the help they can get. If there is an obvious visible signal accompanying the inner feeling—like a bruise—it's straightforward. Not much more difficult, "hunger" corresponds to going for hours without food; "fear" to a large, growling dog or a terrific thunderstorm—generalizing based on these early examples. For more hard-to-get-at inner feelings, metaphors can help, based on (again) generalization. Skinner noted a youngster drinking sparkling water for the first time, who described it as like "my foot's asleep"—presumably because of the shared pinpricking sensations. Only someone who had experienced both sparkling water and a foot falling "asleep" could create this metaphor, but once someone has, other people who have experienced either one can now generalize to the other. In the same way, we feel "sharp" pains or "dampened" spirits. It's not a perfect system, but it's the best we can do for our private sensations.[42]

Our private thoughts can be like silent reading, except we make it up as we go along. Something—anything—starts a train of thoughts and memories. For example, see a product in a store and hear the song that goes with it in TV ads. You might be consciously aware, or you might hear the music automatically even if you don't particularly want to. You might next recall a recent TV show, then the child with whom you watched it.

Cognitive scientists call this "spreading activation." It's related to generalization: when things are associated in some way, thinking of one makes it easier to remember the others. I see a tree that looks like a redwood, and I remember a family vacation to see the redwoods. If I have work to think about, I stop daydreaming and return to it. Competing consequences vie for our private behaviors just as for our public ones.

As Skinner pointed out, simply as part of our language, some words are reinforced when they are presented together as a chain. The first word can "prime" the other, publicly or privately, as in "black and white" (not "black and green") or "shoes and socks" (not "shoes and underwear"). Many members of a culture will share such pairings. Other spreading associations are idiosyncratic. For me, seeing a reference to the nursery tune "Twinkle, Twinkle, Little Star" instantly conjures up an image of my dad and me singing this song, walking to the corner store on a summer evening in Chicago to buy a two-cent pretzel stick. (I would have been about five years old.)

Public or private, thinking often has consequences. (Think about it. Then think, why are you thinking about it?) We generate solutions to problems, for example: How much should that tip be? Indeed, because both public and private rewards come on schedules, it can be hard not to privately obsess over a problem: an addictive variable-ratio schedule keeps us searching for a new angle, a better solution (see chapter 1). From chapter 4, people fitted with the BrainGate Neural Interface System could even move computer cursors with their thoughts.

Thinking about something can elicit the feeling that goes with it (see chapter 8): we smile or frown as we recall happy memories, like mine about my dad, or as we daydream about the rosy future. Thinking can be so like actual events that our mental images serve as signals to which we respond. For example, recalling your recent fender bender, your foot might reach out for the brake. Hash over an embarrassing conversation and you may find yourself talking out loud, saying what you had really meant to say. Indeed, the skin is not that important as a boundary.

Why do we think? For the same kinds of reasons as any other behavior—and that's often because of the consequences.

MAKING UP THE RULES

As one of the consequences of our sophisticated language, our children won't have to reinvent any wheels (we hope). In the course of human history, we have developed rules about what appears to work and what doesn't. Directly or indirectly, short-term or long-term, these rules describe consequences: "A penny saved is a penny earned." "Build above the flood zone." "Don't drink and drive."

When we learn by trial and error, we feel the positive and negative consequences directly. Following a rule can minimize the negatives, and that is part of the reward for creating rules in the first place. But rules themselves are followed because of consequences, new ones as well as the old trial-and-error ones. Obey the queen or lose your head, for example. Governments, bosses, teachers, and parents wield powerful consequences, enforcing the rules (although competing consequences regularly subvert them).

We get so heavily reinforced for sticking to these rules—and punished for breaking them—that we frequently follow them when they are wrong. Even when we *know* they are wrong. If your boss says 2 + 2 = 5, it's best to go along. Rules from fallen idols may no longer be followed, but off-the-wall rules from a trustworthy guide might be. We have been rewarded for following the guide's rules in the past, after all.

Having a rule means we might not notice when things change: rule-following can become so automatic that we forget to pay attention. In one study, rules provided to research participants were accurate at first, but then became inaccurate. The participants eventually realized they needed to ignore the rules, but it took quite a while.[43] We find that we need to learn rules about when *not* to follow rules.

Rules are ordinarily so helpful that we're constantly looking for patterns, making up our own rules. (There are rules to help us do this.) A few striking instances might convince us, or a gradual accumulation of hard-won experience, or careful examination of the objective evidence. Or we might just take a wild guess. And we are frequently wrong. In an experiment on schedules of reinforcement, one participant kept pressing a button when no rewards ever came (royally screwing up the study). Why would anyone ignore consequences so blatantly? From the researcher: "He said he thought that maybe I was trying to see if he was a persistent type of person, one who would continue trying even though success seemed unlikely, and he wanted to show me that he was exactly that type of person."[44]

Larger-scale rules can be unfounded, too, such as superstitious ones. Some hotel chains never have a thirteenth floor; some airlines skip row thirteen (and row seventeen, because that's an unlucky number elsewhere). In Asia, some buildings lack *any* floor numbers that contain a 4. The consequences of these rules for these businesses are clear.

Being a boss is different from having a boss, however current you might be on business-book rules about leadership. Imagine two pilots in the cockpit of an airliner, one more experienced, the other better up on the myriads of rules. Who will act quickly and correctly if the aircraft suddenly loses an engine and plunges to starboard? Why not try for both theory *and* experience? Even then, there is always more to learn. In general aviation (in contrast to the commercial airlines), some studies find that experience appears to have little or no effect on the rate of major accidents.[45] The various safety boards are trying to create new rules to address the problem.

There are rules for everything, including rules about how rules work. After all, rules represent our best attempts to find order in the universe. No wonder our lives revolve around them and the consequences they help us obtain or avoid.

LANGUAGE AND BIOLOGY

All this order relies on our brain with its billions of neurons. A lot of them help support our thinking and language.

We now know that language relies on many parts of the brain. Consistent with the brain plasticity we saw in chapter 4, neuroscientist Phil Lieberman noted that different areas of the cortex can control similar aspects of language in different people.[46] In addition, the primitive basal ganglia are surprisingly critical, even for higher functions like comprehension. As cognitive scientists Elizabeth Bates and Judith Goodman concluded, "Language is a new machine built out of old parts."[47]

Because of their especially high degree of brain plasticity, kids with damage in the brain parts that typically help support language can still learn and use language normally. In adults, for example, damage to the brain's left side often brings language problems. In children, that's not the case: on average, language effects after left-side damage are no different than those after right-side damage. What's more, children who have suffered head injuries that cause language difficulties often recover, given sufficient time and retraining.[48] Then there is the well-known case of Alex, an eight-year-old whose language was severely delayed because of seizures. After his left brain hemisphere was removed—the whole hemisphere—he not only recovered but he achieved normal language skills.[49] Sometimes adults also show surprising plasticity, but nothing to match this.

Finally, a note about the *FOXP2* gene, which has received much attention. It plays a role in human language but also, relatedly, in birdsong and bat echolocation. (It is in fact found in a large number of species.) Like most proteins, the one coded for by *FOXP2* has multiple functions, including support for heart and lung processes.[50]

Whatever future research discovers, language—no more and no less than other behaviors—is the result of a big, complex system full of many interacting factors. Genes and consequences are both included.

PART 3

SHAPING DESTINIES

Chapter 11

EVERYDAY CONSEQUENCES

A female lead printer bore with a lot of putdowns on the shop floor. *"Sometimes I'll compliment myself by saying to one of the guys there, a salesman, 'Don't you think that looks really good today?' And he'll go, 'Oh, yeah.' I need to do something because I get no appreciation."*
—from *Alone in a Crowd*, by Jean Schroedel, 1985

A friend of mine composed a saxophone piece for a semi-pro musician who had expressed interest. When she received no reaction from him, she did not compose again for a year. Screenwriter Rod Serling had better luck. He never forgot an early editor's compliment when his ego had been "bent, cracked and pushed into the ground."[1] One expert recommends keeping a "praise notebook" when necessary.[2] Contemporary jazz goes one better by mandating applause after each solo.

One of life's ironies is that, while we love getting positive reinforcers, we are stingy about giving them out. Yet life is hard; sometimes it seems the world runs on fear. Avoiding negatives provides a lot of motivation, but what a way to live. Philosopher Stanley Cavell wrote of the "little deaths," the everyday cruelties, sorrows, and missed opportunities that drag us down.[3]

Little reinforcers combat the little deaths and help us keep going: A smile. The color purple. Seeing the sky (not a given if your workplace lacks a window, so appreciate it). A word of praise.

If you take nothing more away from this book, I hope it's this: Few of us give or receive enough praise. Yet something so simple can save a marriage, or just help someone get through a rough day.

CREATING REWARDS

Ah, those rough days. Our daily negatives are created by bosses, teachers, parents, spouses, kids, other drivers, uncooperative inanimate objects, and our own mistakes. But we don't have to suffer in silence.

In addition to appreciating the little reinforcers, we can try to transform negatives into positives—like Tom Sawyer's fence painting, singing at work, and the photocopying "flow" I shared previously. Maybe we can't transform our mistakes into prizes, but reminding ourselves to learn from them can help: "At least I'll never do *that* again." Rationalization reframes negatives in what can be useful ways—sometimes.

Here are some more coping techniques. When I'm overcommitted, I schedule tentative events, then I don't go, creating rewarding "found time." I also take advantage of getting a kick out of completing something—anything— relishing the enormous satisfaction of finishing something in a world where there's always too much to do.

We can use what we know about learning from consequences to create or change reward value. Do you hate spicy foods, but your coworkers love them? If so, be prepared to spend substantial periods of time searching Indian lunch menus for bland alternatives. Wimp out with ultra-mild samosas—or consider trying "shaping" (discussed in detail later in this chapter) to gradually train your taste buds.

Our everyday preferences are variable enough, after all. If someone asks what we're in the mood for, we have to stop and think. Even big reversals happen: one-track workaholics fall in love, reformed arachnophobes get pet tarantulas, neatniks become pack rats. Our loved ones keep track and play to these changing motivations, sometimes without even realizing it. The daughter who hates waste finishes the old grungy leftovers because the rest of the family knows she will. The neatnik ends up doing the cleaning (good reason for becoming a pack rat).

Our near and dear also actively bring about changes in consequence value. Telling your husband you're not sure he can handle something immediately creates a reward for doing it; meeting the challenge becomes an accomplishment. More straightforwardly, friends recommend those movies we've got to see, songs we've got to hear.

Because advertisers cannot rely on recommendations, they have to work hard to make their products and services reinforcing enough to try, thereby showing us why they are valuable, better than the competition's, endorsed by celebrities or scientists, or associated with appealing colors and ditties. When all else fails, they may resort to providing free samples. If only such effort went into creating a range of healthy rewards for everyone. The famed explorer Captain James Cook had trouble getting his sailors to eat sauerkraut to help prevent scurvy. He set an example by featuring it at the captain's table and let word get around. It became so popular, he soon had to limit the portions.[4] Rule: if the elite eat it, it must be tasty.

Sometimes creating rewards is almost ridiculously easy. As we have seen, boredom is a state of deprivation that can make a reward out of almost anything. In his book *I'm A Stranger Here Myself*, humorist Bill Bryson noted how silly roadside attractions, like a well-publicized house made of beer bottles, made driving a long, dull highway less painful.[5] Anything to mark progress (see chapter 7), something to look forward to, helps ease the monotony.

What if there are no beer-bottle houses? Kids get cabin fever, but adults are old hands at creating reinforcers to compensate. I recall a long summer road trip in the desert as a kid, before cars had air conditioners. My parents described how they met and got married, more fully than I had heard before. It was riveting, and I forgot all about the heat and boredom.

Although boredom tends to be obvious, other reward deprivations are not. A string of cloudy days brings everyone outside when the sun finally reappears, but we may not even realize we're deprived until then. Deprivation helps create social rewards, too, such that going without attention for just a short while can make even a brief exchange more valuable. Psychologists Jack Gewirtz and Don Baer had more than 100 first- and second-graders play a game that required the children to drop marbles into holes. For children who had first been deprived of attention for twenty minutes, an adult's rewards for marble dropping—words of praise like "Good"—were significantly more effective. These children were also more likely to start social interactions.[6]

Unfortunately, social deprivation—loneliness—can make us vulnerable to the unscrupulous. The reward value of any kind of contact was discussed by one telemarketing victim who lost his life savings: "I loved getting those calls. Since my wife passed away, I don't have many people to talk with."[7]

Simply by asking "What time is it?" a stranger establishes the rewarding value of a reply. The question also signals that the reply will be rewarded in return, if only by attention. On a larger scale, creating curiosity creates reinforcers (see chapter 1). Start a new hobby and see. Or fashion a more ambitious life goal—a dream to work toward.

HOW WE TREAT EACH OTHER

Ambitious dreams of world peace are still unrealized. As we saw in chapter 6, provocation can automatically make retaliation reinforcing, since it can end the provocation. The principle applies to societies and individuals alike.

Encouragingly, though, anthropologists have long known of dozens of cultures in which violence of any sort is uncommon. How is this possible? Among the contributing factors, the Paliya of India and the Mbuti of the Congo carefully pass along their peaceful methods to the next generation (although patterns may now be changing). For example, when a teased child starts crying, Mbuti youngsters are taught to support the victim and exclude the teaser. Nonviolent ways of managing conflicts are taught, modeled, and rewarded.[8]

Clever laboratory studies have shown how our responses to provocation can be influenced by consequences. For example, adults were supposedly playing a computer game with an unseen partner, which was actually just a computer program. The supposed partner gave both rewards and mild shocks, and the participants could respond as they wished. Given shocks, men were more likely than women to give shocks in return, mirroring cultural expectations.[9] In the United States, studies show that aggressiveness is sometimes rewarded in boys and punished in girls (although, given anonymity, women and men tend to be equally aggressive).[10] These tendencies are readily modifiable: when the computer game rewarded aggressiveness in the women, they became aggressive; when it rewarded forgiveness and cooperation in the men, they became friendly.[11]

Much like the Paliya and Mbuti, our own cultures routinely reward aggressiveness or cooperation. Around the world, many of the most powerful rules in everyday lives are such unwritten cultural ones. The consequences of breaking these social norms, even trivial ones, can be surprisingly painful. Famed social psychologist Stanley Milgram asked students to request someone's seat in the New York subway even though other seats were available. The surprised riders proved compliant (what would you do?), but the only student who followed through hated making this request. A disbelieving Milgram attempted it himself: "Finally, after several unsuccessful tries, I went up to a passenger and choked out the request, 'Excuse me sir, may I have your seat?' A moment of stark anomic panic overcame me. But the man got right up and gave me the seat. A second blow was yet to come. Taking the man's seat, I was overwhelmed by the need to behave in a way that would justify my request. My head sank between my knees, and I could feel my face blanching. I was not role-playing."[12] We care more than we would expect about what total strangers think of us.

ALTRUISM

Here is one of the more surprising findings from social psychology. If you have an emergency, you might think it best to have four or five strangers nearby rather than just one. But you might be wrong: researchers consistently found that people in a group were *less* likely to help than a sole individual. One of the reasons was "diffusion of responsibility," such that no one took charge in the group—there being no consequences for doing so. When you are the only one around, it's up to you, the consequences are quite different, and bystanders become more likely to step forward.[13] Because of this consistent finding, some people trained in CPR are instructed to direct one person out of a group of bystanders to call for help, and not to simply ask the group. "Bystander education" programs are also becoming popular in schools, with or without any accompanying discussion of altruism.

According to John Steinbeck's jaundiced view in *Travels with Charley*: "A man who seeing his mother starving to death on a path kicks her in the stomach to clear the way, will cheerfully devote several hours of his time giving wrong directions to a total stranger who claims to be lost."[14] The status and control inherent in giving directions can be less-than-selfless rewards, it's true. Fortunately, we're capable of more altruism than that. In cultures that emphasize cooperation, mutual assistance can become an automatic habit that is rewarding to boot.

Altruism means helping others for few or no extrinsic rewards, even at a cost. Feeling another's pain is a direct way for relieving it to be intrinsically rewarding: if you suffer the vicarious pain, you also enjoy the relief. It's not surprising that research supports a role for such empathy: in one study that used physiological measures, participants who felt more for a suffering victim were also more willing to help.[15] As we saw in chapter 8 on classical conditioning, viewing someone else's pain can produce brain-activation patterns characteristic of our own pain. Husbands were given shocks while their wives watched, for example. The wives' brains registered pain on an fMRI, and the ratings of the experience correlated with their brain responses. Other research has shown that psychopaths lack the normal degree of our merciful ability to empathize.[16] Less empathy results in less altruism (and how).

Seeing the result isn't necessary. The imagined satisfaction of a person who has been helped can become rewarding, with altruism as a natural result even with no public recognition to assist. The intrinsic satisfactions of following ethical or moral rules also chime in, and we can feel like heroes, if only to ourselves. It's possible to derive great satisfaction from potential, imagined changes in the future that we won't even live to see.

Altruism isn't all sweetness and light, though. Anyone who has felt bitter regret knows how powerful a motivator avoiding it can be ("I could have saved her"). Sometimes it's the consequences for *not* helping that loom large: how can you live with yourself or deal with the disappointment of the people whose opinions you value if they find out? It's yet another example of multiple causation.[17]

One way or the other, we can learn to enjoy helping others. Caring can become rewarding at an early age, and once it's rewarding, it naturally leads to action. Children are often praised for showing sympathy, and of course they see it modeled and they read about it in children's books.[18] Gradually they learn emotional signals and the beginnings of perspective taking.

SHAPING THE FUTURE

As parents, we reward altruistic behaviors ("How nice of you to share, Johnny") and we try to set a good example. But it's a long "shaping" process, like so much of parenting.

Simple forms of shaping occur in birds such as spotted flycatchers raising their young. The nestlings learn that parents signal food, and when the chicks can fly, they start chasing mom and dad. The parents reward this, but soon start feeding on a steadily leaner intermittent schedule of reinforcement rather than every time. Having learned what's good to eat, the young naturally start catching their own food. The consequences "shape" a steady move toward independence by rewarding successive approximations to this goal.[19] Wild meerkats have a similar system for teaching how to catch and eat poisonous scorpions. In their society, as in ours, the teachers aren't necessarily family members.[20]

Shaping can start early: newborns of many species can learn from consequences,[21] and human infants are no exception. How do babies tell us what's rewarding? They suck artificial nipples. In one research procedure, two signals alternate. If a baby sucks during one signal, one sound is played; sucking during the other signal causes the second sound to play. If the baby doesn't suck at all, no sounds play. It's straightforward to show that effective rewards include mom's voice over another woman's; mom's native language over a different language; and even a passage that mom had read repeatedly before birth, compared to a different passage.[22] All this occurs when babies are only one to three days old.

Soon babies turn their heads, smile, and do other neat things because of rewards.[23] We saw in chapter 10 that babbling babies as young as two months produced either more consonants or more vowels, following positive reinforcement. Karen Pryor reported a fishing captain who shaped his four-month-old granddaughter to "Gimme five!"[24] Shaping helps sitters become crawlers, and crawlers become toddlers.

Two-year-olds working on the alphabet make lots of mistakes at first, but we reward their steadily closer approximations to the correct twenty-six letters in the correct order. Learning the alphabet is not intrinsically rewarding, nor does it offer the natural rewards of, say, being able to speak a foreign language. Because that's the case, parents offer attention, smiles, praise—artificial or "contrived" rewards, if you like. (Some experts consider these to be "natural" rewards.) Unlike learning the alphabet, reading does offer natural rewards. However, a parent encouraging a reluctant reader might again offer praise, with the idea of ultimately transferring to the natural rewards. They are part of the shaping process, too.

The idea of natural consequences applies to negatives as well. Parenting expert Thomas Phelan gave a true example of a preschooler who procrastinated instead of getting dressed on time in the morning. The boy's frustrated mom finally let him be taken to preschool in his pajamas. The natural embarrassing consequences meant this was never a problem again.[25]

Phelan and Yale professor Alan Kazdin, past president of the American Psychological Association, provide useful lists of suggested artificial rewards in their parenting books, including staying up late and points on a point chart, exchangeable for privileges or other rewards.[26] Are these helpful or even necessary? Consider this: For some lucky few adults, their work is so naturally rewarding that they would do it for free (baseball star Babe Ruth comes to mind). But for most of us, artificial paychecks are an essential incentive, although of course we appreciate any natural rewards that come with the job. The fact is that very few kids find learning multiplication tables intrinsically rewarding.

With either natural or artificial rewards, shaping works for a broad range of behaviors. Kids shaping a teacher into speaking louder will pay attention and smile whenever his voice rises—at first only slightly, then more, until he's close to a shout. The same approach works just as well for speaking more quietly or moving closer to the window. On a larger scale, shaping builds behaviors of far more significance: social skills, study habits, assertiveness, conflict resolution—indeed, even parenting itself. Guidelines for shapers include not raising the bar too quickly.[27] It's not hard to understand why. If a three-year-old makes progress on a difficult, intrinsically unrewarding task and gets absolutely nothing for it, for example, why should he bother?

Shaping is so important in learning that roboticists developing artificial intelligence are carefully studying the science.[28] Despite that unusual application, ordinary real-time shaping is as much creative art as science, making for a highly enjoyable parlor game or classroom demonstration.[29] Try taking turns as "shaper" and "shapee." Pick a mystery behavior, then use a clap to reward successive approximations to switching on the light, say, or raising both hands high. My students loved it.

In his autobiography, B. F. Skinner discusses how he shaped psychologist Erich Fromm's hand motions as they talked, by paying more attention, nodding, and smiling, and got vigorous waving in response. It worked fine and without Fromm realizing what was going on.[30]

Because human shaping is not necessarily planned or even in our conscious awareness, however, we can get blindsided into shaping the opposite of what we want. Guess what happens next time if parents give in to a tantrum? Waiting it out is best, but "extinction bursting" means the tantrums sometimes get worse at first. If parents eventually relent, they have just shaped *more extreme* tantrums, and the kids have learned that if they ramp up, they will be rewarded.

On that subject, is there hope for violent teens and for the rest of us who have to cope with them? In observing thousands of parent-child interactions, child-rearing expert Gerald Patterson saw these negative patterns get shaped.[31] Children readily learned to escalate their protests to get what they wanted if that approach paid off (recall the aggressive four-year-olds of chapter 7). Discipline in these families was typically inconsistent, and appropriate praise was lacking. Programs to help parents switch to a more positive, consistent style have resulted in significant benefits, including reductions in vandalism, truancy, gang membership, dropout rate, and substance use, and increases in achievement test scores.[32] Patterson's own parent-training program met the criteria to be considered "well-established" scientifically in reducing later violence and aggression.[33]

Positive does not mean constant compliments, by the way: in fact, praise can lose its value that way and become unrewarding. Anything gets cheapened if it's too easy to get (see chapter 5) or if it's insincere. Real appreciation has real value.

While it's more natural for us to notice problems like fighting than non-problems, desirable behavior cannot just be taken for granted. Parents have to learn to see kids behaving well and then let them know. As authors Adele Faber and Elaine Mazlish noted in their bestselling book *How to Talk so Kids Will Listen and Listen so Kids Will Talk*: "Most of us are quick to criticize and slow to praise. We have a responsibility as parents to reverse this order."[34] Catch 'em being good.

Unfortunately, fighting over a toy can be naturally rewarding: the winner gets the toy. Can shaping with contrived positive reinforcement really work for playing peacefully? Five-year-old Martha was mean and disruptive and (understandably) avoided by the other kids in her preschool. The solution? First she was shaped just to interact with them and then to play with them. That is, her teachers provided attention and occasional praise whenever Martha spoke to the other kids, then when she played with them appropriately—successive approximations to the target of cooperative, peaceful play. Finally, when the preschool teachers smiled and attended to her only for playing cooperatively, Martha responded by doing it more.[35] (When the same social consequences were presented independently of her behavior, there was no effect on her cooperative play, so they were indeed rewards for it.)

Once a child like Martha starts playing without fighting, the natural, intrinsic reinforcers of suddenly having friends often take over. It's one small step toward peace.

THE CHALLENGING SIDE OF PARENTING

For serious behavior problems like fighting, though, sometimes rewarding desirable behavior isn't enough, and talking it over doesn't always work either (especially with young children). Timeouts—formally invented by consequence researchers many years ago (see chapter 6)—have their place. Both the Paliya and the Mbuti cultures use timeouts and discourage corporal punishment, the same policy adopted by the American Academy of Pediatrics.[36] Extensive research shows that even brief timeouts from positive reinforcement can be an effective alternative to more drastic punishment.[37] While timeouts can help reduce a problem behavior, they do not teach desirable behaviors, of course, so the emphasis should continue to be "catch 'em being good."

Of the many different forms of timeout, the most restrictive is removal from the activity to a separate room. At the other end of the spectrum, the child stays in the activity, but participation is stopped briefly, or a favorite toy is briefly removed. The TV program *Supernanny* (ABC) relies heavily on positive reinforcement and the "naughty chair," an intermediate form of timeout.

Does timeout always work? Consultant Ennio Cipani related the tale of a four-year-old child who was put in a timeout location right next to a toy shelf. He simply began playing with the toys.[38] For any form of timeout to work, timein has to be rewarding and timeout has to be substantially less rewarding. Ideally, the minimum duration that is effective would be used, usually quite short.

In one study, four-year-old Rorey was disobedient and aggressive, punching and slapping other children. In the intervention, his mother ignored mild problem behaviors, while more serious ones brought removal to a timeout room that had no playthings. Mom stated immediately but quietly why Rorey was going into timeout, which lasted two minutes. A tantrum during timeout meant another two minutes from the end of the tantrum. (Otherwise, children learn that screaming will get them out of timeout.) Meanwhile, Mom rewarded appropriate play and instruction-following with praise and occasional special treats. Before timeouts, Mom suffered through as many as thirteen aggressive incidents per day. After just two timeout days, Rorey's problem behaviors almost disappeared, and timeout was rarely used.[39]

What about older children? In *How to Talk So Kids Will Listen and Listen So*

Kids Will Talk, Faber and Mazlish discussed the helpfulness of mutual problem solving and natural consequences in avoiding emotional tussles and contrived punishers. A teenager who borrowed a sweater from his dad gave it back covered with chalk and spaghetti sauce. The natural consequences included dad's reluctance to lend out clothing again, which made sense to the teen. The next time he wanted to borrow from his dad's wardrobe, the boy wrote a note promising a return in good condition—and lived up to his promise.[40]

But sometimes negotiations break down, positive methods fail, and all heck breaks loose. Alan Kazdin and colleague Carlo Rotella wrote an online post subtitled "What to Do When Your Kid Provokes You into an Inhuman Rage." They suggested that one good response is based on having discussed ahead of time the loss of privileges like TV watching. Given a child's outburst, impose the penalty, calmly explain why, and walk away. Kazdin calls this "the parking ticket" approach. As always, it should be accompanied by positives for good behavior, rules consistently enforced, and exemplary role modeling (as much as possible; kids imitate their parents).[41]

Best of all, we can head off problems before they start. Karen Pryor discussed how to deal with kids who are habitually too noisy in the car. Reward an incompatible behavior: For example, start a game like "I spy" before the kids get wound up. Shape gradually longer periods of appropriate noise levels with a favorite snack or the chance to play a handheld game. Or change the motivation, thus changing the consequence value. Enforced inactivity makes rowdy behavior more rewarding. The kids might not get antsy if you stop regularly so they can run around.[42]

You might even be able to pull a Tom Sawyer and transform a negative into a positive. One at-home dad told his story in Faber and Mazlish's parenting book: He created "a new way to deal with all the mess the kids make. I take out my special deck of cards with all the high numbers removed. Then each boy picks a card that tells him how many things he has to put away. There's lots of excitement as they count what they put away and rush back to see what their next card will be. The last time I did it, the whole cleanup was finished in twenty minutes and the kids were disappointed that the game was over."[43]

WHAT MARRIAGE CAN BE

For ourselves just as for our kids, it's important to accentuate the positive. Something so simple can indeed save a marriage (but that doesn't mean it's easy).

Marriage researcher John Gottman first popularized the so-called "magic" 5:1 positives-to-negatives ratio after observing many typical husband-wife conversations. A good predictor of divorce was too many negatives, not enough positives—not just words, but nonverbal signals like a blaming or supportive tone. Gottman's research showed that some couples exchanged lots of positives for every negative, exceeding 5:1 by a substantial margin. Others languished at *less than* 1, exchanging more negatives than positives. Consistently low ratios of positives to negatives were clearly a danger sign, but how high was high enough? Gottman picked 5:1 as a good proportion to shoot for.[44]

It has become the "magic 5:1 rule," now popularized in a number of books, including Tom Rath and Donald Clifton's bestseller *How Full Is Your Bucket?*[45] It has also expanded well beyond marriage. We've seen how valuable a high positives-to-negatives ratio appears to be in predicting language success (as discussed in chapter 10), and its benefits are noted in the parenting books I've mentioned along with many others. A number of research studies now support it in marriage, in child-rearing, and in additional areas like education, business, and prison rehabilitation.[46] Researcher Barbara Fredrickson found that a minimum 3:1 ratio enhanced emotional resilience (the ability to handle aversives), and noted that most of us fall short.[47] For everyday purposes, the exact ratio probably isn't important, but the idea is.

Just as for our kids, high positive ratios make it safer to experiment, establishing an atmosphere of trust. Every couple argues, but Gottman found that was no problem if the overall positives-to-negatives ratio was reasonably high. In happy marriages, negatives were approached confidently as occasional coping transactions, because positives were the norm. (A couple could disagree about what to do over the weekend, all the while aware that some sort of compromise was bound to be reachable.) It may be that in almost any human—or animal—relationship, the same basic principle applies. We certainly know the harmful side effects of too many negatives (see chapter 6).

Journalist Amy Sutherland took a consequence-based approach to her own marriage. While better than most, it still had its share of annoyances (for example, dirty clothes left on the floor). Talking sometimes helped, but sometimes not. When it didn't and Sutherland resorted to nagging, that often made things worse. No one likes being nagged, but it's reinforced on a variable schedule, which keeps it going. Down plummets the positives-to-negatives ratio.

Sutherland stopped nagging her husband to put his laundry in the hamper. Instead, she ignored the piles lying around, and simply expressed her gratitude whenever he did toss something in. Gradually, he improved (shaping). To keep him away when she was cooking, she didn't nag, she created tasks elsewhere in the kitchen such as setting the table (rewarding incompatible behaviors). This positive approach worked so well that her husband adopted it too. The marriage, already strong, improved noticeably.[48]

How about starting from scratch? In conversations, one Joe Schmoe with poor social skills failed to listen or ask questions, sticking interminably to his own favorite topics. No wonder he was rarely successful at getting dates. At his request, women trained in shaping social interactions rewarded desirable conversational sallies and ignored or mildly punished (discouraged) boring ones. He improved so much that he married one of his practice dates within the year.[49] His self-esteem got a boost too.

REAL SELF-ESTEEM

Should we be focusing on our kids' self-esteem? Psychologist Roy Baumeister, who had been a strong proponent of doing so, reviewed the extensive research literature. Surprisingly, he concluded that in itself, self-esteem boosts like "what a great kid you are" simply didn't help, producing no benefits.[50] It was specific praise for real effort and accomplishment that did the trick—and of course that kind of praise is far more likely to be a real reward that actually influences what kids do, along with how they feel.

Just letting youngsters know that their efforts *can* make a difference seems a straightforward way of motivating and empowering them. But does it work? Self-esteem researcher Carol Dweck taught study skills to underprivileged junior high kids. Half the students were also given information about how effort can improve intelligence, even reading an article about how learning changes and enhances the brain (as discussed in chapter 4). Kids in this group did significantly better that semester, even though the teachers did not know who had been in which group (so no self-fulfilling expectations could have confounded the results; see chapter 14).[51] In other research, Dweck found that kids praised for their intelligence were less likely to attempt challenges and were more likely to give up when they failed.[52]

Just as praise can be empty and unrewarding, so can attempted self-esteem boosts that are not supported by the empowering relationship between effort and its consequence of progress toward achievement. It seems that kids benefit from learning from their mistakes along the way, too. Lacking experience with mistakes and other natural negative consequences, kids may have more trouble coping with them later. (Hovering "helicopter" parents, take note.)

In the same way, the sooner, the better when it comes to learning how to handle those frustrating delays to the consequences we want. Self-control may be the most valuable skill that parents can nurture.

Chapter 12

FIGHTING THE IMPULSE: SELF-CONTROL, ANYONE?

"The hardest victory is the victory over self."

—Aristotle

T
here's a reason why self-control has been glorified for millennia: it's hard to achieve. Most rewards don't happen when we want them to, which is right away. Witness the temptation to struggle with debt rather than wait and save for a big purchase.

Societally as well as individually, get-yours-now is hard to resist. The Atlantic Grand Banks cod fishery, one of the richest in the world for centuries, crashed in 1992 from overfishing. Some experts fear it will never recover. Unfortunately for us and for our planet, our susceptibility to short-term rewards can bring long-term disaster.

Fortunately, we can use what we know about consequences to combat our worst impulses. This chapter focuses on the individual challenge, chapter 16 on the societal one.

DETECTING DELAYS

Two researchers occasionally rewarded rats with food for poking their noses into a board with lots of holes. What hole would they return to after they had enjoyed their treat? It wasn't the one they had poked ten minutes ago or even one minute ago. It was the one they were poking when the food came.[1] The logic goes: do something and get an immediate "consequence," and perhaps what you did caused it. It's worth checking out, anyway. It's like the superstition effect (see chapter 5): Bump against an electric pole right before a blackout and fear that you caused it. Bump half an hour before, no worries.

In nature, this emphasis on immediacy usually makes sense. Move your legs faster and as a consequence you move faster. Right away, not ten minutes later. Catch a fly by flipping out your tongue. Immediately.

Still, ambush hunters like cats learn to wait outside a hole for a tasty mouse. In nature and in the lab, animals can handle some delays. B. F. Skinner found that rats new to the game could pick up lever pressing with an eight-second delay to the reward.[2] Later research stretched this out: Plop Daisy the rat into a box with a lever. With nothing else happening, Daisy eventually presses the lever (hey, it's something to do). Thirty seconds later, a food pellet is delivered. Meanwhile, Daisy's doing other things: sniffing, turning a circle, licking a paw. If she repeats these, she gets no food. Eventually, she presses the lever again. Thirty seconds later, another food pellet. That's a long delay for an animal, but rats can eventually get the connection.[3]

How do we know it's the consequence that starts them pressing regularly and not some other effect of the food? Because when they get the same amounts of food at comparable, variable times but independent of what they do, they *don't* press the lever. Even Siamese fighting fish can learn real but delayed rewards in this way.[4]

Consistency and immediacy are both influential, then, but neither is absolutely essential. One immediate consequence reinforced the nose poking of rats. But so did a consistent dependency without immediacy, as with Daisy's thirty-second delay.

Not surprisingly, people learn new behaviors on this delayed reward procedure too.[5] With enough other things going on, though, it becomes hard even for us to detect the real relation between the consequence and what we did. Consequences, signals, and everything else operate simultaneously over different periods of time. What a muddle. No wonder we have trouble being influenced by delayed consequences: sometimes we can't even tell they are there.

Language bridges delays with helpful rules: Use sunscreen. Take your umbrella. Even long delays in ancient times: plant almond seeds now, harvest almonds years later.

But even with the big advantage of language, look how long it took for us to recognize what caused scurvy. No vitamin C; hello, scurvy. But the consequences are delayed, and the variety of suitable foods to prevent scurvy quite varied. Decades after British naval surgeon James Lind finally proved the effectiveness of fresh citrus fruits, the British navy adopted this approach—but the loss of vitamin C caused by poorly preserved juice meant that British sailors and explorers continued to suffer from scurvy. Fresh meat alone sometimes worked, so fruits and vegetables clearly weren't essential, adding to the confusion. For a while, it was even thought that scurvy might be caused by tainted meat, possibly helping to doom Captain Scott's fatal South Pole expedition.[6] Disentangling real but delayed consequences from the red herrings is one of the triumphs of scientific method.

THE DISAPPEARING REWARD

Now turn this on its head and see how quickly the value of a consequence drops the longer it is delayed. That makes it hard to weigh properly. And that's where self-control comes in.

Our lives are constant choices between consequences (see chapter 7), and many of those consequences are delayed. Eat too much, become overweight. (Later.) Study hard, land a good job. (Later.)

Then there's uncertainty: some delayed consequences are on lean schedules with low probability. For example, don't buckle up, risk more injuries in an accident. (Later. Maybe.) It took years for safety-belt use to become the norm, despite overwhelming evidence of its benefits.[7]

What's more, the same self-control choices can work exactly the opposite for different people. For most of us, for example, saving rather than spending is the challenge. But the stringently frugal have to force themselves to spend—taking self-control too far. The same principle applies to the tragedy of suicide. If felt to be justified and morally necessary, suicide can be a supreme act of self-control: the pain and finality of death is exchanged for the longer-term good of others. Or it can be a supreme failure of self-control: escape overwhelming immediate pain, but miss out on all sorts of delayed reasons to live.

Every day, we ordinary heroes choose present pain for future gain, larger-later desirables over smaller-sooner less desirables (see chapter 6). We take a walk rather than watch TV, turn down the temperature to help the planet and save on the heating bill later, and get chores done instead of procrastinating. Or at least we try. The parallels for animals remain surprisingly close. In a study called "Procrastination by Pigeons," the birds chose more work later over less work now.[8] Sound familiar?

Not surprisingly, self-control has attracted lots of research, and that has enabled scientists to mathematically describe the way in which delayed consequences lose their value. My colleague George Ainslie was one of the first to show that the relationship follows a "hyperbolic" curve, an equation that works for a variety of consequences in a variety of species and across large as well as small time scales. It often works well even when people choose between hypothetical rewards (such as money) to be received at different times, knowing that they won't actually receive them.[9]

Quite apart from the obvious benefits of rules and reasoning, there are exceptions to this "delay discounting."[10] For example, knowing you will be publicly humiliated tomorrow spreads gloom today, and you might well prefer to get it over with. So much for the more typical procrastination effect of putting negatives off, with larger-later preferred over smaller-sooner. As in this case, sometimes classical conditioning helps make the waiting period itself a negative, so smaller-sooner gets chosen instead. For positives, it's usually the impulsive smaller-sooner over larger-later. But, like happy memories in reverse (see chapter 8), the pleasures of anticipation can override the usual delayed drop in a reward's value. One economist asked men to place values on when they could—hypothetically—kiss their favorite movie star. The highest-valued was the three-day delay, not the immediate embrace.[11] There is a reason for songs about anticipation.

What's more, while impulsiveness is usually deplored, sometimes it really is better to take the sure thing now and be done with it. When Poland suffered hyperinflation in the early 1990s, its people learned to spend quickly because their money actually did lose value overnight. When the economy stabilized, this impulsiveness disappeared.[12]

Most tests of delayed consequences fit the hyperbolic equation well, though, and it succeeds in predicting a peculiar feature of self-control: we change our minds as time progresses and consequence value alters. We start with good intentions: wake up, work out, and resolve to skip dessert at dinner. But hours later, staring at a brownie, we decide maybe we can afford those extra calories after all. Similarly, setting the alarm the night before is easy, but getting up when the alarm goes off isn't. What about hypothetical consequences? Imagine a choice of taking $100 now or $150 in a year. How about $100 in two years versus $150 in three years? Many people reverse their preference.

Animals do the same: given a choice between two seconds of eating delicious grain now versus four seconds of eating in four seconds, pigeons go for smaller-sooner. Add ten-second delays to both choices, though, and the birds get rational: they select larger-later. Same reversal. Same equation.[13] When rats and people both work for real juice available after different delays, their results are quite similar.[14] Rats consistently show more self-control than pigeons when it comes to food.[15] But with learned rewards rather than food, pigeons fall closer into line.[16] And just like us, individuals can be somewhat more or less impulsive.

THE MARSHMALLOW AND THE KID

How about kids? From psychologist Walter Mischel's famous research line, picture a four-year-old left alone to choose between one marshmallow and two. It's no contest—except that getting two means waiting fifteen minutes. With one tempting marshmallow left sitting in front of them, very few kids hold out: some simply eat it immediately. Those who waited substantially longer did substantially better on their SAT college entrance tests many years later.[17] A 1,000-kid study in New Zealand had similar findings, showing benefits twenty-five years later for those with more self-control as youngsters, even taking into account other factors like socioeconomic status.[18]

In one version of Mischel's study, some kids tried to make the negative of waiting into a positive. "They talked to themselves, sang, invented games with their hands and feet, and even tried to fall asleep while waiting—as one child successfully did."[19] We saw in chapter 5 that two-and-a-half-year-old children on a low-speed schedule developed similar tactics to make it through the delays—pacing around the room, for example. So did some animals.

So how would pigeons do on the marshmallow challenge? Experimental psychologist Allen Neuringer did the research to find out. Birds that waited got the delicious grain they preferred. Birds that gave in to temptation pecked a key and took a smaller amount sooner—and what's more, it was a kind of grain they found boring. Kids did best when they couldn't see the marshmallows. So did the pigeons. Like Mischel, Neuringer also looked at what happened when an alternative activity was available, when the birds were given successful experience with waiting, and when signals helped. Again, results were similar.[20]

For yet more similarities, it's long been known that people who test out as more impulsive in these sorts of lab situations tend to have more problems with smoking, drinking, and drug addiction.[21] This relationship applies in rats too, surprising as that may seem. Rats that were more impulsive on a standard delayed-food test were more likely to give themselves cocaine and to use more of it.[22] Of course, these relations tell us nothing about what causes what—a classic correlation-doesn't-equal-causation muddle. (Other research shows that rats given cocaine *become* more impulsive—and remain so even several months after they have been off the drug.[23])

Fortunately, there is no question that self-control can be taught. We have already seen that monkeys reared by adults were less impulsive than those reared by other young monkeys (see chapter 3). Mischel studied several methods, as noted previously in the current chapter. We humans have found many ways to nurture self-control.

One approach is very straightforward: researchers gave kids an immediate choice between big and small rewards, then gradually delayed the big reward. Would the kids learn that waiting was worthwhile? Yes, and all the more impressive because these particular children had been selected by their pre-school teachers as being especially impulsive.[24] In his bestselling book *Walden Two*, Skinner speculated that his fictional community might teach this critical skill in just this sort of way.[25]

This method can work for animals, too. In one study, rats chose between two levers providing large or small amounts of food, both delayed by six seconds. They picked the large amount, naturally. When the delay for the small reward was then eliminated, the rats continued preferring the delayed large reward (in contrast to what they would have done without this experience). Next, when the two levers reversed in function ("smaller-sooner" became "larger-later" and vice versa), all but one of the original twelve animals reversed also, staying with larger-later, the wiser choice. Impressively, most of them kept showing self-control for delays of up to twenty-four seconds, quite a lot for a small animal.[26]

FIGHTING THE IMPULSE: USING WHAT WE KNOW

News we all can use: examples of consequence-based self-control assistance.

Adding/subtracting consequences. Our fast-food, sound-bite world constantly serves up instant gratification. Witness the texting mania: walking while texting, driving while texting, even skydiving while texting (seriously).

Modern technology can make self-control easier by adding immediate rewards for it. Many vehicles now provide instant mileage feedback, for example, particularly efficient hybrids. No more waiting until you fill your tank to figure out your fuel efficiency. As a result, a new subculture has sprung up, and "hyper-milers" compete for super efficiencies. One hybrid owner referred to the change in driving style as "a little addictive."[27] Who would have thought self-control could be fun?

Using the same strategy, some utilities now offer inside-the-home electric meters displaying real-time usage, sometimes including rewarding color changes for conservation. A pilot program in Canada showed a significant 13 percent average decrease in consumption once customers were able to watch how much power they used.[28]

If there are no natural immediate rewards, try adding artificial extras. A psychologist "would reward herself by playing the video game 'Asteroids' after she had met each daily writing goal. 'By the time I had finished writing my dissertation, I had well over a million points.'"[29]

Let's not forget the dark side. Add negatives: For instance, dab your fingernails with a bitter solution so you don't bite them. Watch scary ads showing sick smokers breathing out of tubes in their throats.

A target of saving, say, $1,000 creates a useful additional reinforcer and beats a loose wish to spend less. Similarly, having a deadline creates a reinforcer for meeting it (see chapter 6). Can homegrown deadlines work? A study by behavioral economist Dan Ariely and a colleague found that students did better with self-created deadlines than with none, but deadlines imposed by others were the most effective.[30] Be careful, though, not to lose sight of the forest for the trees. Researchers find that people who miss their weight-loss goals often report feeling like failures even if they had succeeded in losing some weight.[31] Effective behavioral shaping can take time.

Social support and models. Simply letting others know your goal means extra rewards—and punishers. Encouragement or disappointment from other people can be a powerful motivator. Partner with a friend to exercise or study. Sharing your efforts means learning through observing each other, too.

"Women in Red Racers," an online debt-reduction support group, uses a number of these strategies. Post your debt and your progress toward paying it off—or your backsliding. Everyone in the group can see and encourage you, as in well-known self-monitoring groups like Weight Watchers and Alcoholics Anonymous. From one of the Racers' members, "having a forum where you can say, 'I just got a $25 birthday check, and I'm putting it toward my Visa balance'—and then getting a round of cheers from your fellow Racers—it's incredible." A system of rewards for milestones includes a smiley face for each $100 off your debt. In less than two years, a few hundred members together paid off about $3 million. "You wouldn't think smiley faces could be so important," said a member.[32]

Signals. Some natural signals come from our own bodies. Detect when you're starting to stress, and you may be able to head off an angry outburst.

Artificial self-control signals are common, too. There is a good reason that radio DJs announce upcoming hits: they signal that staying tuned through a bunch of ads will be rewarded. Relatedly, have you noticed that when you have a full tank of gas, it's harder to exert self-control about unnecessary trips? The next visit to the gas station is too delayed to be effective. When you're getting low on gas, it's easier to drive less and conserve fuel. Signals matter.

We saw that in chains of behavior, each behavior signals the next. These chains often become automatic. Get in the car, buckle up; you don't even think about it. Toothbrushing is a classic example of an even more tight-knit chain.[33] Some chains end in trouble: a TV commercial starts and you immediately wander to the kitchen to get a snack. Train yourself to jog around your living room instead. Break the chain.

Finally, just as for the kids eyeing the marshmallows, anything that draws attention to smaller-sooner rewards can hurt. Hide your temptations. For some, that might be a favorite video; for others, cigarettes or alcohol.

Schedules. Like all rewards, the Women in Red Racers' smileys were earned on a schedule. Some self-control methods build directly on their different qualities.

Chapter 5 introduced a schedule of reinforcement in which progressively more behavior was required over time—toddling, for example. Rewards came only with longer distances toddled. My colleague Steve Higgins introduced a related schedule for drug abstinence, a massive self-control problem. In this schedule, the behavior of interest doesn't change, but the reward value does.

Cocaine addicts had their urine tested three times a week. Being cocaine-free earned a voucher exchangeable for specific items such as clothing, movie tickets, or stereo equipment. The vouchers got steadily larger as the addicts succeeded in staying clean. Fail a test or fail to report in, and the voucher size reset down to its initial value.[34] Voucher programs have proven to be among the most effective in drug treatment (see chapter 15 for more).

A common self-control challenge is getting stuck with a grueling, lean fixed-work schedule (that is, a long ratio to work through for each reward). You can suffer through the characteristic long pause before starting work or schedule extra reinforcers to help get yourself moving. Even better, switch from a fixed to a variable reward schedule. On the variable "reading-the-newspaper" or "gambling" schedule, a reward might always be right around the corner. Steady work is what's typical—but only if the variable schedule includes occasional easy rewards.[35] That's how important they are. So schedule them. (Las Vegas does.)

Commitment. Remember the reversal in preference as time goes by? Make a commitment far enough in advance that you choose wisely and avoid later temptation. Some sleepyheads put their alarm clocks on the other side of the room so they can't just hit the snooze button. At the grocery store, never buy your favorite brand of potato chips—less chance later of instant gratification at home (not shopping when you're hungry helps, too). In several states, problem gamblers can voluntarily ban themselves from casinos and their "addictive"-like variable schedules. If the gamblers go anyway, they face fines and even arrest. Thousands have put themselves on the lists.

It may seem unlikely, but animals can also learn to make commitments and thus maximize their rewards. Psychologist Ainslie gave his impulsive pigeons the option of pecking a commitment key before each choice. If they did so, then after a delay only the larger-later option would be available, no smaller-sooner possibility. Three of the ten birds tried it and learned to commit reliably.[36] Another researcher found that requiring multiple pecks on either key rather than just one choice peck led to more larger-later choices (technically, a "fixed ratio"). The extra delay in doing the extra pecking—even though it was short—was enough to jump-start self-control. Switching in the middle of the extra pecks meant starting them from scratch on the other choice, and a lot of wasted effort for the bird. The birds learned to commit to their first and wiser choice.[37]

Behavioral economists Richard Thaler and Shlomo Benartzi developed the Save More Tomorrow (SMarT) program to help people overcome the difficulties of making careful financial choices that have delayed consequences (see chapter 1). With an opt-in default for pension plans, for example, employees are automatically entered in a plan. Far more participation results, which is beneficial for the employees.[38] Making a commitment ahead of time eases the way toward wise financial management.

Checklists and charts. Simply checking items off a list, keeping a progress chart, or creating a monthly budget and sticking to it can be surprisingly rewarding. One writer on a deadline noted, "I kept track of my progress on a wall calendar, with a gold star for each week I met my goal. At the end of twenty-six weeks, I had twenty-six gold stars and forty completed chapters. This may sound corny, but it really works."[39] Plenty of electronic "apps" are now using these techniques for self-control challenges as well as a variety of other motivational purposes (even dating or starting a hobby).

Skinner went further, charting exactly how much time he spent writing (see chapter 7), and Anthony Trollope and Ernest Hemingway used a similar strategy, tracking how much they wrote.[40] It works for smokers, too. Karen Pryor graphed the number of cigarettes she smoked each day and was able to quit—for good. Seeing her overall progress helped her through her lapses.[41]

Experience. We've seen that simple practice on a schedule of gradually increasing delays can teach self-control, letting the natural reinforcers be experienced. Similarly, mystery novels are universal favorites: Whodunit? Good writers build the suspense, so the reward value of finding out gets large enough that some of their eager readers turn right to the end. But by persevering through the subtle cues, the delayed reinforcement is all the larger, and most of us learn to exert self-control—with experience.

Catherine the Great described how she learned from experience: "I have made it a rule to begin always with the most difficult, most awkward and most tedious matters; with that out of the way the rest seems easy and agreeable."[42] It's like scheduling dessert at the end of a meal.

Rules. The more we learn, the more sophisticated and effective the rules we try to live up to. Many of us count to ten or leave the room when we're angry, following age-old self-control rules. Author S. E. Hinton suffered from writer's block for four years after her first book, *The Outsiders*, became a sensation. Her boyfriend finally suggested that she make it a rule to write several pages each day—or he wouldn't take her out in the evening. It worked.[43]

Following rules can be rewarding in itself. Even though I have been working evenings and weekends for years, I still get a virtuous glow because of the cultural rule that I'm going above and beyond. (No wonder I used to be a workaholic. Everything has its dangers.)

Because of cultural rules, seemingly straightforward self-control incentives can be anything but. Two behavioral economists investigated late pickups at daycare centers. When parents started paying a $3 fine for being more than ten minutes late, substantially *more* came late, overall. Why? Because before, parents felt guilty about delaying the staff. Now being late was just part of a minor financial transaction.[44]

TAKING CHARGE OF WEIGHT

The need for self-control has never been greater when it comes to maintaining a healthy weight. According to the Centers for Disease Control and Prevention, two-thirds of Americans are now overweight, and one-third qualify as obese.[45] This epidemic has developed just in the past twenty-five years. The costs are astronomical. Consequence-based strategies can help.

Extra rewards. Over time, the natural reinforcers of feeling better and losing weight may be all you need to exercise and eat right. But artificial rewards can boost motivation. I work out with free weights at home twice a week, for example, but I used to make excuses to myself and skip some sessions. Now I read for fun during the short rest periods between sets and I perform this workout more reliably.

More extra rewards. New recipes for healthy foods can make them taste better, and that can make a big difference. (I've come to love vegetarian cookbooks.)

Positive signals. Some dieters post refrigerator photos of themselves when they were thinner. Negative signals can work, too: dieters leave the room when scrumptious TV restaurant ads come on. Yes, watching them can make you feel hungry (see chapter 8).

Charting/models. For youngsters up to age eleven, the award-winning Food Dudes program features healthy-eating cartoon role models and a flexible variety of rewards and progress charts for trying fruits and vegetables. Artificial rewards are discontinued when the natural rewards take over. Enough research has documented the program's success that Ireland now uses it in all its elementary schools.[46]

Charting. From expert Miriam Nelson, "Study after study has shown that if you record your progress in a fitness program, you're much more likely to be successful."[47] Likewise, one study found that, while instructions alone increased the amount of swimming that children on a team did, recording their progress at the end of each session was more successful.[48]

Schedules. In chapter 5, a boy getting physical therapy exercised more on a variable-ratio schedule—"an enjoyable game"—than an equivalent fixed ratio. (This should no longer be surprising.) Similarly, obese boys exercised more on a variable schedule than on a comparable fixed one.[49] Easy, this one.

Here is an example that may be less easy to follow: one of my friends and colleagues tried progressive schedules to help him exercise more. It's akin to shaping: he started at twenty pool lengths per swim and increased that number by 10 percent with each new session until he was up to a mile— seventy-two lengths. Surprisingly, he described the experience as "painless." No doubt this can be easier than it sounds, but do try to keep your target and your step-up rate realistic.

Commitment/social support. New apps like the free weight-loss program "Lose It!" (www.loseit.com) take a positive approach, including commitment and social support, checklists and charts, and a variety of "gamification"-type rewards.

Finally, Yale behavioral economist Ian Ayres cofounded stickK.com to help people with their self-control challenges. Any goal is legit: learning a new language, quitting smoking, spending time with family. The most popular is losing weight.

The stickK behavioral contracting system relies more on negatives than positives. For example, you can put up money that will be donated if you fail to reach a weekly weight goal. The funds can go to your grandma, a favorite charity, or an anti-charity—a charity of whose goals you disapprove. StickK simply manages the transactions. For accountability, an agreed-upon referee can check whether you meet your goals. For social support, you can list friends, relatives, and coworkers who will be informed of your progress—and your failures.

It's a potent combination. At this writing, in less than four years stickK handled over 100,000 contracts and more than $9 million, and dieters lost many thousands of pounds. Not surprisingly, Ayres reported that contractors were more successful when they put up more money, utilized a referee, and had more supporters cheering them on.[50]

Ayres himself used the site to drop some weight, putting up $500 each week to be donated if he hadn't lost at least one pound. It worked so well that he didn't lose a dime. By continuing on a weight-maintenance contract, he has successfully maintained a healthy weight, thereby avoiding the yo-yo weight cycling that bedevils so many dieters.[51]

Victory over self just got a little less hard.

Chapter 13

ENDANGERED SPECIES, UNDERCOVER CROWS, AND THE FAMILY DOG: APPLICATIONS FOR ANIMALS

> *"Near the southern tip of Brazil, a cooperative fishing method has arisen. . . . During our visits typically 30–40 fishermen and one to four dolphins were present in the principal fishing location throughout the daylight hours. . . . Town records state that the cooperative fishing began in 1847. Some fishermen report that their fathers and grandfathers fished before them, sometimes with the same individual dolphins. . . . Fishing does not begin until a dolphin initiates it."*
> —Karen Pryor et al., *Marine Mammal Science*, 1990

A dolphin ready to fish swims away from the beach, then turns toward shore, driving mullet toward the half-immersed fishermen standing in the water. They usually can't see the fish, so they toss their nets only upon observing the arriving dolphin roll in a distinct way. The nets drive the fish that escape back toward the dolphin, a win-win situation for everyone except the fish. Dolphin mothers bring their calves, and the young appear to learn from consequences through observation and practice.

There's something magical about this sort of cooperation between wild animals and us, communicating across the species barrier, helping each other.

Our knowledge of how consequences work has helped us help animals around the world. Large zoo animals no longer need to be anesthetized for routine medical attention, for example. Even bears cooperate willingly through positive-reinforcement training, supported by the Association of Zoos and Aquariums in its Animal Care Manuals. Further, behavioral enrichment through consequences means no more pacing tigers bored out of their skulls.

These methods are helping endangered species recover. They are improving conditions for farm animals. And they are reaching the multitudes of animals that help *us*. (Guide dogs are just the beginning.)

The most widespread use of our knowledge of consequences, though, is in our own homes with the animals that are our daily companions. Toss those choke collars: positive-reinforcement-based methods for pets are backed by science, taught by many veterinary schools, endorsed by the Humane Society of the United States,[1] and popularized by the long-running shows *Calling All Pets* (NPR) and *It's Me or the Dog* (Animal Planet; sort of like *Supernanny* for Spot).

ANIMAL COMPANIONS

People domesticated animals many millennia ago. When did some of them become pets? We'll never know. But the United States now hosts roughly 80 million pet dogs, 90 million pet cats, and plenty of other pets.[2] Ferrets, fish, finches . . . we love them all.

The rewards of companionship from a different species aren't limited to humans. Wild badgers and coyotes have been documented to hunt semi-cooperatively—and to play together.[3] Racehorses get attached to their mascots: goats, cats, even chickens. In an endangered-species preserve in Myrtle Beach, South Carolina, an orangutan befriended a stray dog, playing with him, sitting with his arm around him, and sharing food.[4] Other orangutans have had cats as "pets."[5] Perhaps even more startling, an adult pigeon in China adopted an orphaned rhesus monkey youngster of about the same size, and they snuggled together.[6] The mutual consequences that keep these relationships going work for us and our pets, too.

Remember the benefits of a high positives-to-negatives ratio (see chapter

11)? Using what we know about consequences means emphasizing positive reinforcement with pets as well as with people. In one such approach, pair a clicker's abrupt sound with powerful rewards to make it a learned reward. Then click right after a behavior you're teaching. Why a clicker? Simply, it's a convenient immediate marker or "bridge," followed reliably by a reward of some sort. After you train the behavior, train a signal: Do the behavior when the signal is given, click. Do it otherwise, no click. The basics are simple.

In her bestselling classic, *Don't Shoot the Dog!*, pioneer Karen Pryor described teaching clicker training to new trainers. In twenty-four hours, one ambitious novice taught her "clueless" shelter puppy "sit, down, roll over, come, a super 'high five' in which the little puppy rolled its weight to the left and threw its right paw straight up as far as it could reach into the air— and the beginnings of a retrieve. All on cue, rapid-fire, correct, and in any order. The puppy, furthermore, was electrified, a totally different dog, attentive, full of fun, muscles all engaged—ready for life." And she concluded, "any creature—a dog, a horse, a polar bear, even a fish—that you shape with positive reinforcers and a marker signal becomes playful, intelligent, curious, and interested in you."[7] I clicker trained my parakeet Goldie to do a somersault around her perch. And one of my friends clicker trained her cat.

A few standard pointers may be helpful here. Just as for people, it is best to use many smaller rewards to get a behavior going, and then to maintain it on a variable schedule. Give out bigger rewards for more difficult behaviors. And shape approximations to what you want, like the students shaping their teacher to speak louder (see chapter 11). Playing the "shaping game" in that chapter helps us appreciate how animals feel when we are shaping *them*. "Is that what you want? How about this? I get it!" It's fun. Reward creativity and dolphins go wild with ingenuity (see chapter 1); later researchers found that this works for dogs and pigeons, too. As Pryor put it, positive reinforcement gives an animal "its own magnificent ability to Make People Do Stuff."[8] The rewards are mutual; the sense of communication, powerful.

Once behaviors and signals are established, you don't need the clicker any more—regular rewards keep the behavior going. Examples from Pryor include "a pat or a smile, or just a chance to do other good things such as going for a walk."[9] As biopsychologist Ray Coppinger noted, "A good retriever sits there and begs you to throw the ball again. (Actually, I've often thought that my dog should be rewarding me with a biscuit. . . .)"[10]

This doesn't mean that problem behaviors won't crop up. Teaching an incompatible behavior can help: reward your dog for greeting guests by waving a paw instead of jumping on them. A clever way to handle problems is training a signal for the behavior you don't want—begging for food in the kitchen, say—but then not presenting the signal much. In other words, you reward begging only when the signal is present, then gradually stop giving that signal. It's counterintuitive, but it really can work. If necessary, timeouts are used as mild negatives during training. Because training is reinforcing, a pause means that fewer rewards are available. Animals notice.

Many shelters support the move away from stronger negatives, sometimes offering dog harnesses that don't hurt in exchange for choke, prong, or shock collars.[11] (Shock collars are banned in a number of nations, including Germany, but not in the United States.) Training that relies on punishment can mean fewer voluntary behaviors (why take a chance?) and more stress. One traditional trainer was dismayed when her dog hid under the porch at training time.[12]

In *Reaching the Animal Mind*, Karen Pryor described the start of an experiment comparing clicker training to traditional training, using several dog trainers who had "crossed over" after careers in the traditional approach. Getting untrained shelter dogs to lie down on command traditionally doesn't mean using choke collars, but just gently pushing the animals down. Even so, the trainers were now uncomfortable seeing the confusion and stress the dogs were feeling, something they knew was easily avoidable with clicker training. The experiment stopped almost as soon as it started.[13]

From a different crossover trainer who had seen this sort of stress: "I witness the diametrical opposite when the clicker comes out. The dogs are frantic to train. They are falling all over themselves, eyes glowing and wide with anticipation. They are offering all of the behaviors in their repertoire just to get a chance to get their turn to train. . . . There is no comparison whatsoever in the results, contrasting the traditionally trained dog with one trained with all-positive reinforcement. And, there is no going back."[14]

AT THE ZOO: ANIMAL CARE THE EASY WAY

Talking about the move to positives, Ken Ramirez, vice president of Chicago's Shedd Aquarium, won't even let his trainers say "No." If they use reprimands, he says, they will eventually overdo it.[15] In the bad old days at the zoo, though, negatives of all sorts were the norm. Simple animal care required capture, drugging, or restraint. (Visualize *Crocodile Hunter* Steve Irwin immobilizing a rogue reptile; it's necessary but stressful.) Fire hoses sometimes forced large animals to move when their cages needed cleaning, and few exhibits offered much in the way of behavioral opportunities. Now animals help out willingly for rewards on cue, and positive-reinforcement training and behavioral enrichment are standard. All sorts of species have benefited.

Even painful animal care can be provided positively by taking advantage of schedules of reinforcement. Animals used to getting fake needle sticks for a reward can handle a real blood test once in a while. A typical study compared levels of the stress hormone cortisol in monkeys giving blood, either when they were pursued, caught, and restrained (the traditional way) or cooperatively presenting a leg upon request (after an hour of positive-reinforcement training). Far lower cortisol levels occurred in the reinforcement condition.[16] Keepers enjoy this approach a lot more, too.

Relatedly, after only two to three minutes of daily training for nine days, three African wild dogs learned simple behaviors like holding out a paw, which is helpful for administering medical care. For an hour after each training session, their stereotyped pacing was a fraction of what it was at the same time on non-training days.[17] Sea lions shared similar benefits.[18]

Positive-reinforcement training impacts other problem behaviors as well. At the Bronx Zoo, animals learned basics like going on a scale or inside a crate in response to a gesture. Clicker training helped most animals catch on in just a few short sessions. As a result, primates from tamarins to saki monkeys stopped retreating in fear when a keeper came near. Instead, they "eagerly approach and interact with the keepers and voluntarily participate in the training sessions."[19] At Disney's Animal Kingdom Theme Park, "the benefits to animal care and welfare have been enormous."[20]

As a special type of signal, "targeting" means an animal comes to an object like a hand, a pole, or the dot from a laser pointer and follows it wherever it goes. It comes in handy for pets and zoo animals alike, and search-and-rescue dogs follow laser pointers into hard-to-reach areas. Move zoo animals to a different area simply by moving the target—no more fire hoses. And training can be fast even for reptiles; an endangered loggerhead sea turtle learned targeting in just a week.[21]

Gestures can work, too. At Walt Disney World, spotted eagle rays were bothering divers. After training, divers being buzzed by rays pointed to a diver with a feeder and the rays headed there, for all the world like undulating spaniels showing off in a field trial.[22]

To understand why problem behaviors like buzzing occur, a "functional analysis" (see chapter 9) is just the ticket. The reasons aren't always obvious. A baboon at a zoo started plucking her hair, and the functional analysis showed no itchy skin disease, just human attention rewarding and maintaining this problem behavior. The ape was successfully taught to smack her lips for attention instead, and the hair plucking stopped.[23] Attention may well have been rewarding the rays' buzzing too.

LIFE AT THE ZOO

In the wild, animals have full lives. What do they do in a zoo? Animals that get bored may seek attention through good means as well as bad. From chapter 9, recall the orangutan that was taught to imitate; she went on to initiate mimicking games with zoogoers to the pleasure of all.

But we can do better. Behavioral enrichment, based in part on the science of consequences, is now a standard at most zoos. Zoo expert Kathy Carlstead and colleagues describe a naturalistic approach: "An environment in which an animal can find food as a consequence of its natural exploration and foraging behavior is an essential key to approximating natural habitats and to improving animal welfare."[24] Even simple approximations help. As these researchers found, when Asian leopard cats had to search for hidden food, they showed fewer unhealthy stereotyped behaviors, just like the black bear described in chapter 1. A "bug toss" at one aviary was appreciated by many of its free-flying birds.[25] And a cougar kept "hunting" an artificial ground squirrel even when no more food was provided for a successful capture.[26]

Enrichment pioneer Hal Markowitz is known for letting animals initiate and control these activities.[27] After all, control can in itself be a reinforcer (recall the deer mice from chapter 1). And it can reduce stress: for example, rhesus monkeys given control over whether a loud noise was on or off showed no more stress (based on cortisol levels) than monkeys that didn't hear any noise. Without control, the monkeys became stressed and aggressive.[28] We can relate.

Useful enrichment doesn't have to be naturalistic. Offering inherently reinforcing variability as well as control, a computer game at one zoo could be played any time. Many apes and monkeys like video games (see chapter 10), and a male mandrill that took full advantage became less aggressive, which was a big relief to the members of his troop.[29] Other monkeys enjoyed just turning a radio on and off, listening for hours each day.[30]

Toys producing unpredictable movements can be big hits (like the suspended bag for the rhino in chapter 1). Similarly, SeaWorld animals learn a variety of behaviors—as many as 200—to keep healthy and stimulated.[31] In addition, SeaWorld relies on reinforcer variety: recall the stroking, toys, attention, and other unpredictable rewards for their orcas described in chapter 1.

What about those bored, pacing tigers? At the Bronx Zoo's Tiger Mountain, a large naturalistic environment, the big cats can do much of what they would normally do in the wild: hide behind shrubs, scratch fallen trees, even swim. Tigers rotate among different areas for more variety, plus the chance to enjoy novel smells donated by the public. Positive-reinforcement training provides for their medical care. Seventy-five different enrichment items include a ball-like "treat spinner" puzzle and a "tiger fishing pole" that lets the cats reach out for deer hides. There used to be the chance to play tug of war with the public, so rewarding that keepers sometimes had to step in to get the tigers to let go of the rope. The endangered big cats appear healthy and happy, and produced two sets of triplets recently. From a human viewpoint, visitors named Tiger Mountain as their favorite exhibit.[32]

FROM ENDANGERED SPECIES TO FARM ANIMALS

One of the most important functions of zoos and aquariums these days is the conservation and captive breeding of threatened and endangered species. Around the world, zoos have cooperative programs, exchanging breeding animals for genetic diversity and sharing information about what works. Positive-reinforcement training and behavioral enrichment are critical.

Members of an endangered species of ibis, for example, were not doing well at one zoo. When positive-reinforcement training was introduced, the birds took to it immediately—even seniors over twenty years old. Not only was their medical care easier but they also lost their longstanding fear of people, and they even started to play.[33] That bodes well for captive breeding.

Once you've got youngsters, how do you prepare them for the big scary world out there? One threat to the endangered California condor is electrocution from power lines (not something this ancient species had to worry about until recently). Because these birds have wingspans of up to ten feet, they are especially likely to complete a circuit and electrocute themselves accidentally. With only about 400 in the world at this writing (and only half of those in the wild), each individual is precious. Scientists now teach captive-reared young to avoid power lines by punishing them with shocks when they land on them. The trained birds have been far less likely to run into trouble with the electric grid after they're flying free.[34]

Back from the brink, the white-winged guan of Peru had been considered extinct for almost a century. Rediscovered in the 1970s, only about 300 survived in 2012. Captive breeding has been augmented by antipredator training of the young that relies on learning through observation. Trained hawks hunt and capture chickens within the sight and hearing of the young guans. This is drastic but effective training for the guans, unlike alternative approaches. When the hawks are then simply flown over their pens, the youngsters try to flee—just the response they need in the wild. Without this learning experience, they are sitting ducks. As it is, some reintroduced birds managed to breed during their first year in the wild.[35]

Innovative consequence-based techniques have also been applied to conserve species while they are still roaming their native habitats. In an all-too-common scenario, when outside species invade an ecosystem, they can destroy the balance of nature. In Hawaii, rosy wolf snails were intentionally introduced to feed on an edible African snail that had also been intentionally introduced. Unfortunately, the African snail had turned out to be an agricultural pest on the loose—and farmers complained. But no one told the wolf snails to dine only on their intended prey, and they targeted the native Hawaiian snails too, driving a number of them to extinction. Using positive reinforcement, Working Dogs for Conservation trained two eager recruits to sniff out the wolf snails, and they successfully found hundreds for removal.[36]

Such expert sniffers have helped the conservation effort in many ways. For example, a controlled study showed that trained dogs reliably located threatened desert tortoises, significantly outperforming human teams.[37] And spaniels have tracked near-threatened ornate box turtles at the Upper Mississippi River National Wildlife and Fish Refuge, gently fetching them for radio tagging.[38] It's a start toward their conservation.

Farm animals have received their share of attention, too. They may not have the crowd appeal of pandas or penguins, but ordinary pigs quickly learned individual feeding-time signals, enabling specialized diets to be provided more easily. What's more, with a variable button-pushing schedule at the trough, pigs enjoyed some of the benefits of the comparable control-it-yourself approaches at the zoo.[39]

What about chickens? Enriching their cages with a simple string reduced neighbor pecking.[40] Other researchers showed that hens liked chicken wire flooring more than a type that had been thought to be preferred (by people).[41]

Our mathematical understanding of choice between consequences means animals can rate their preferences with precision. Chapter 7 described how a New Zealand research group used the matching law to precisely quantify cows' preferences for different feeds. This sort of research has also let milk cows tell us what suction levels were most comfortable.[42] Because being heavy with milk is a drag, the chance to be relieved is a reinforcer. Studies have demonstrated the merits of cow-run milking systems: each bovine determines when she wants to be milked and simply goes when ready to an automated milking machine, several times a day. In some cases, special food treats increase the reward value of being milked.[43] These cow-run systems are now commonplace (which was news to me).

Indeed, the capabilities of farm animals shouldn't be underestimated. Sheep are often handled with trained dogs, but they are quite capable of learning signals themselves. When an outsider blew a sheepdog whistle at one herd of dog-free sheep, they obediently quit grazing and walked toward a nearby gate.[44] Portugese expert Fernando Silva clicker trained a lamb to target, step in a box, follow a pointing command, and do standard agility stunts, including jumps, ramps, teeter-totters, and tubes.[45] (See the YouTube video and do a double take—is this a dog? The internet URL is in the notes section.)

Finally, five quarter horses hated getting into horse trailers because of the whipping and roping they had suffered. By the end of positive-reinforcement training, though, given the signal to load, all five voluntarily walked up the ramp all the way in and waited for the door to be closed. All five also generalized their ability to a new trainer and trailer.[46]

ANIMALS THAT SAVE OUR LIVES

Think of animals that help us and a guide dog may well come to mind. While rewards were (of course) always part of canine education at the seventy-year-old Guide Dogs for the Blind, the recent addition of clicker training significantly increased the program's success rate.[47] Clicker training also clicked for a famous assistance animal, "Dog of the Millennium" Endal, subject of a recent Hollywood movie. Partnered with a Gulf War veteran who was confined to a wheelchair and suffering from memory loss, Endal reportedly learned well over 100 spoken and signed commands. After an accident that knocked his partner to the ground, "Endal immediately put [his partner] in the recovery position, covered him with a blanket from the wheelchair and moved the mobile phone close to his mouth. Then he alerted staff at a nearby hotel."[48] Wow. Assistance dogs have many lifesaving feats to their credit.

Lifesaving dogs also serve in search and rescue, explosives detection, and, surprisingly, sniffing out disease. Who would have thought that cancers can have characteristic odors? Stomach cancer, colon cancer, breast cancer, prostate cancer—all relatively common, all killers—are all detectable by dogs. Trained dogs have managed to beat some of the standard diagnostics. In one controlled study, a female black Lab was 97 percent accurate at spotting colon cancer based on stool samples, far better than the inexpensive fecal blood test that's commonly used. Her reward? Playing with a ball.[49] Other studies have confirmed these successes.

Another way to save lives is to find the millions of hidden, unexploded landmines around the world. Dogs can handle this too, at just a small risk of setting them off. One group of researchers is clicker training dogs to detect land mines more safely just from air samples.[50]

Rodents have good noses too, and giant pouched African rats are helping to de-mine Mozambique. Because they are much lighter in weight than dogs, they never set off the mines. Alan Poling, an expert in the science of consequences, helped develop a system of clicker training that rewarded successful finds on a variable schedule.[51] This meant that the rats persisted in their search for long periods even with no mines to be found. Again, the trained animals became very accurate: in one study, all thirty-four rats succeeded in becoming licensed for mine detection. That means they sniffed out all of the five to seven mines hidden in an area. Also important, along the way they gave no more than two "false alarms"—indications of mines where there were none. In 2010, the "HeroRAT" teams located more than 1,200 mines and "unexploded ordinance" left over from Mozambique's civil war,[52] and they are now supported by heavyweight donors like the United Nations and the World Bank.[53]

What's next? War itself. Trained dolphins have served the US military for years, detecting marine mines with their sensitive sonar so that experts can remove them. Reportedly, "in 2003, at the outset of the Iraq War, dolphins were brought in to clear mines from the Umm Qasr harbor to allow a humanitarian relief ship to enter."[54] A 2010 drill in the San Francisco area included antiterrorism work by Atlantic bottlenose dolphins and California sea lions.

In a famous example from World War II, the Soviet Union trained thousands of dogs to attack invading German tanks. At first, dogs were supposed to drop a bomb and return, but this approach didn't work. Instead, dogs carried bombs that exploded on impact, killing the dogs, but helping to save Allied soldiers.[55] The project was never very successful, but got further than a comparable project in the United States. Meanwhile, thousands of homing pigeons served as wartime messengers, and the United Kingdom considered (but rejected) using trained pigeons to carry small explosives.[56]

The pigeon project may well have worked. From an actual over-the-sea search-and-rescue simulation: 93 percent detection accuracy for trained pigeons peering down from a Coast Guard helicopter, only 38 percent accuracy for competing humans.[57] So don't dismiss the odd prospect of undercover crows seeking terrorist leaders in hiding, reportedly a recent military project.[58]

Researchers have found that wild birds like crows and pigeons can distinguish and remember human faces (even mockingbirds—remember chapter 7), and they aren't fooled by a change of clothes.[59] Given their visual acuity and outstanding memories, maybe this is not so outlandish an idea after all.

Indeed, to end this chapter, we move from undercover crows back to pets: an outlandish-sounding but true tale. Naturalist Edwin Way Teale recounted the story of a pet crow that had a special friendship with a young boy. When the boy went to school, the crow followed, perching on the outside window sill of the room the boy was in. "When the boy moved to another room, the crow flew from window to window, peering in, until it caught sight of its friend. Then it waited patiently on the window sill until another change was made."[60] Now that's companionship.

Chapter 14

THE REWARDS
OF EDUCATION AND WORK

*I want my students to understand that their ability to read and write
is a matter of life and death.*
    *Successful classrooms are run by teachers who have an
unshakable belief that the students can accomplish amazing things
and who* create the expectation that they will.
          —Rafe Esquith, inner-city fifth-grade teacher,
          There Are No Shortcuts, 2004

In the 1960s, psychologists Robert Rosenthal and Lenore Jacobson famously told eighteen elementary schoolteachers that some of their students were likely to begin doing well, based on their scores on an intelligence test. Actually, these students were randomly selected, no different from the others. Surprise: most did indeed show bigger gains on a repeat of the intelligence test at the end of the school year.[1] This basic finding has now been repeated many times.

Clearly, the teachers' expectations had influenced the outcomes, but how? Consider how you might respond if you were a teacher. Perhaps you would give the special students more to do and call on them more? Treat them more positively? (You might not even realize what you were doing.) Rosenthal and another colleague showed from an analysis of 135 of these studies that all these things happen.[2] And those kids benefit.

Quite apart from research on such self-fulfilling expectations, other studies have demonstrated the importance of the changes in what the teachers did (such as giving students more attention and more work to do). The lesson, of course, is to expect the best from all students and use these effective methods with all of them.

That is exactly what Disney National Outstanding Teacher of the Year (1992) Rafe Esquith does.

THERE ARE NO SHORTCUTS

In his inner-city Los Angeles school, Esquith's fifth-grade class meets from 6:30 a.m. to 5:00 p.m., plus some Saturday afternoons: extra time for math and reading, while still allowing time for science, geography, economics, music, and Shakespeare. So he can focus on teaching, not classroom management, Esquith developed a point system (more on these later). Much of class time goes to practice: doing problems and reading. Esquith agrees that self-esteem in itself does little or nothing (see chapter 11); it's real achievement that counts. His students work hard. And they achieve.

They also take field trips. "As a teacher of children from economically disadvantaged backgrounds, I came to understand that my students would work harder for a better life *if they saw the life they were working for*."[3] To make sure the trips would be rewarding, he gave his students lots of preparation. (Before attending a concert, for example, Esquith would have students study and discuss the music.) The idea is to support a transition to the natural rewards of accomplishment. Along the way, ideally, learning can become its own reward. What could be more transformative?

Like Esquith, the late Jaime Escalante was an inspiratonal teacher who required his high school students to put in extra school-day hours, plus some Saturdays. The movie *Stand and Deliver* (1988) cheered the math achievements of his underprivileged Hispanic students, who passed the difficult Advanced Placement Calculus exam at unusually high rates. The students worked intensively on math basics first, saying and doing, not just listening (which is better for learning and memory); took daily quizzes; and engaged constantly in "practice, practice, and more practice."[4]

It's ultimately about motivation, and that means consequences. Escalante pointed out that students could attend college and get better jobs with good math skills—and graduates of his program returned to share their successes. Like Esquith's students, Escalante's took field trips. They visited NASA's Jet Propulsion Laboratory, for example, where one of his former students worked. For more immediate motivation, Escalante as coach and the students as team (wearing team jackets and often studying together) worked toward their common goal. Keep learning rewarding despite the hard work, he said: "Students learn better when they are having a good time."[5] Escalante chose his texts for their intrinsically rewarding real-life examples, their readability, and their programmed, cumulative building of skills. "If motivated properly, any student can learn mathematics."[6] Escalante won the Presidential National Medal of the Arts for excellence in education in 1998.

Most good teachers at any level are unsung heroes. They have high expectations. They use effective teaching methods. And, since students must be good classroom citizens in order to be taught effectively, they develop effective classroom management. At the elementary- and middle-school levels, that can be particularly challenging.

CONSEQUENCES IN CLASSROOM MANAGEMENT

Consequences are, of course, inevitable and unavoidable in any classroom, under any approach. For example, a teacher's mere attention can be neutral, rewarding, or punishing, depending on the situation and the nature of the attention. While praise is rewarding for most youngsters, several studies have shown that it can actually be punishing for students with a history of negative interactions.[7] (In addition, some students get teased or harassed by their peers for their academic success, and that can detract from the consequence value of praise.) Conversely, in another study, second- and third-graders who acted out responded well to soft reprimands, but loud reprimands *increased* their misbehavior—and thus were actually rewards by definition (likely due to the reinforcing attention these students received from their classmates).[8]

Formulating rules in a positive way helps: expect good behavior. When I taught high school briefly in the Peace Corps, students wanted to be there, and I don't recall any classroom-management problems. In many cases, though, rules alone aren't enough. When researchers observed several disruptive kindergarten and second-grade kids in regular classrooms, for example, rules alone had almost no effect. Success: adding occasional praise for specific class behaviors, while ignoring misbehavior.[9] (And once again it goes without saying that effective praise must be sincere.) These mild consequences for the whole class were also easy to do, an important consideration for busy teachers.

A classic study by parenting expert Alan Kazdin (see chapter 11) showed how different students can respond differently to the same classroom management system. In six regular elementary classrooms, results were contrasted for obedient children and for kids with less glowing histories. Different classrooms tried out different combinations: (1) All students were told they would get points for behaving appropriately, and they were indeed given points accordingly. (2) Students got the same points, but with no instructions. (3) Students were given the same instructions, but the points were actually delivered randomly. (4) Random points were given with no instructions.

Given points for appropriate behavior, all children tended to improve whether instructed or not. With no instructions and random points, there was no effect. Finally, when told they would be rewarded for appropriate behavior but actually given random points, normal kids followed the rules anyway. Disruptive kids did not. So it's best to be consistent (no surprise). After the program was over, the classes with true rewards continued to behave better.[10]

Many teachers use such point systems in some form, and they come in all shapes and sizes (and go back at least to the early 1800s). Points, tickets, stars, smileys, tally marks, or fake coins can be exchanged for a variety of backup rewards, so they work for everyone. The forty-year-old "Good Behavior Game" is a popular group point system. Teachers assign teams with roughly equal mixes of well-behaved and "problem" children. During game periods, teams get points if everyone behaves appropriately, but lose points if any member fools around (incurring social consequences). Sometimes all teams can win if they make enough points; sometimes there's a competition.[11] This approach has worked in a culture as different as Sudan's,[12] and kids years later are less likely to be aggressive or to start smoking.[13] In a recent study, kindergarten teachers preferred a positives-only version to the version that includes point deductions.[14]

Like many instructors, though, master teacher Rafe Esquith prefers a combination of positives and negatives, and an individual rather than a group approach. His students have jobs for which they get "paid" in class dollars. In turn, they pay rent for a seat (higher rents for the front), or they can buy a seat and pay property tax. Just as dollars can be earned for good test results, good attendance, extracurricular activities, they can be lost for rudeness, lateness, missing homework, dishonesty (big-time).[15] Education expert Lee Canter's more conventional Assertive Discipline system also relies on both positives and negatives, with points deducted for problem behavior.[16] (Point deduction is generally preferred over a timeout because it's faster, easier, and less disruptive.)

Other systems are negatives-only, but they are not as discouraging as the term makes them sound. Students in one class had their own traffic-signal boards with the arrows set on green. Disruptive behavior meant the arrow moved to yellow; if turned to red, the student would take a note home and lose half of the next day's recess. All students started each day with green.[17]

No system is foolproof, and details matter: "catch 'em being good" is all very well, but how often? In past decades, observers recorded overall positives-to-negatives classroom ratios of 1:2, 1:3, and even lower, a far cry from the 5:1 "magic ratio" recommended for marriage and parenting (see chapter 11).[18] Talk about discouraging.

If the ratio gets turned around, what happens has the potential to deserve the "magic" billing. In one informal study, 80 percent of students in a school in a disadvantaged neighborhood were assigned to special-education classes. The positives-to-negatives ratio in these classrooms was about 1:4, so the researcher asked teachers to find more behaviors to reward. Shaping—a teaching basic—means starting with small steps in the right direction and building from there (see chapter 11). The teachers (and students) responded brilliantly, achieving a flabbergasting 40:1 ratio. The next year, only 11 percent of students were placed in special education. Everyone liked the change.[19] (In a different program, social worker Mark Mattaini movingly noted how novel it was for at-risk kids in "circles of recognition" to be praised. "Within less than two minutes . . . everyone was in tears, sometimes even the group leader, because the experience was so new and powerful."[20])

In a more formal study, the ratio was 1:3 in an elementary school control group, but it reached 4:1 in classes where teachers received training. Inappropriate behavior declined significantly in these classes compared to the control group. Improving the ratio frequently improves academic success as well: better classroom management facilitates teaching and learning.[21] (Query: Is there teaching *without* learning?)

Finally, another classroom-management approach takes advantage of the benefits of variable schedules: teachers reward whoever is working and following class rules when a timer unpredictably goes off. In one classroom of second-graders, on-task behavior increased from 58 percent to 93 percent.[22]

When the timer was discontinued, though, it reverted to an intermediate level of 75 percent. New behavior patterns cannot just be taken for granted and ignored after they have been achieved. Abrupt switches are particularly tough: the natural rewards often take time to gain value. In one study, six seriously disruptive elementary schoolkids in a special class finally responded to a point system with timeouts, and then gradually were transitioned to praise alone with occasional point rewards. It worked.[23] Education researcher Hill Walker suggested that giving unexpected rewards occasionally should be a standard practice—taking advantage of the benefits of variable schedules.[24]

MAXIMIZING POTENTIAL

We have seen that nature and nurture always work together, providing immense flexibility. Once classroom management is in good shape, how far can students go? Esquith's and Escalante's inspiring examples show how much good teaching can accomplish. Can "intelligence" as measured by IQ—a score on a test—be increased through methods like effective teaching? The evidence shows that it can.

It is well established, for example, that simply rewarding disadvantaged children for trying hard on intelligence tests can immediately raise their IQ scores by ten points or more.[25] (Without *some* source of motivation, why strive to do their best?) A recent meta-analysis assessed the findings of many such experiments, including over 2,000 participants altogether—children of all sorts, not just disadvantaged children. Overall, rewarding youngsters for trying harder significantly raised IQ scores, and larger incentives consistently produced larger effects. The effects were greatest when the original IQ scores were lower (not surprising).[26]

Studies suggest that, other things being equal, IQ increases as a direct function of time in school.[27] The "Flynn effect" is the same positive trend on a bigger scale and over a longer period of time: namely, in the twentieth century, IQ scores in the United States rose considerably in a few generations—roughly ten points per generation—because more Americans had been able to get education and proper nutrition. The same thing happened in other developed nations, and, more recently, this effect has been documented in places like Kenya.[28]

Separate research documents the boost that proper nutrition gives to IQ.[29] Then there's the research of psychologists Betty Hart and Todd Risley (see chapter 10): regardless of race, sex, parents' income, and other such factors, children who were given enriched language backgrounds were able to start preschool with a jump on kids who hadn't had these advantages. (And follow-up studies by other researchers have confirmed these findings.[30]) Hart and Risley also found that the more parents talked to their kids, the faster the children's vocabularies grew, and the higher their IQ was at age three.

Building on this theme, psychologist Anders Ericsson, education expert Benjamin Bloom, and others studied the heights of human achievement, from science to the arts to sports. How does expertise develop? Experts didn't necessarily have unusual gifts, the researchers concluded, they just put in the long years of study and effort required to reach the top.[31] Take chess, for example. The longer you study, the more you remember, and that helps your game. Building chess knowledge over the years also means being able to take advantage of memory-boosting techniques like "chunking" (that is, seeing patterns instead of individual chess pieces). That helps your game, too. But chess masters proved to have better memories *only* for chess patterns that could actually exist. For random organizations of the pieces, they did no better than regular players, nor did they have better memories in general.[32] It's the serious, "deliberate" practice that pays off, just as in Escalante's math classes.

While serious practice brings progress toward a goal, it's seldom notice-able enough to be intrinsically reinforcing. How do budding experts acquire the motivation to persevere? According to Ericsson's research, they get it from the prospect of long-term rewards like fame and fortune; short-term rewards like performances, contests, and praise; and progressive schedules with gradu-ally increasing practice time, which develops the ability to handle lean sched-ules of reinforcement with long delays (see chapter 5).[33] Like the kids with the marshmallows (see chapter 12), they exercise self-control, but the marshmal-lows are a lot bigger. Bloom came to similar conclusions, noting in addition a role for the positives-to-negatives ratio when discussing the deeds of the teachers of these experts when they were young children: "These teachers gave much positive reinforcement and only rarely were they critical of the child. However, they did set standards and expected the child to make progress, although this was largely done with approval and praise."[34]

One of my favorite quotes on maximizing potential is from David Shenk's book, *The Genius in All of Us*: "Everyone is born with differences, and some with unique advantages for certain tasks. But no one is genetically designed into greatness and few are biologically restricted from attaining it."[35] Another is from Malcom Gladwell's *Outliers*: "To build a better world we need to replace the patchwork of lucky breaks and arbitrary advantages that today determine success . . . with a society that provides opportunities for all. . . . The world could be so much richer than the world we have settled for."[36] Evidence-based education is attempting to do just that.

SUCCESSFUL PROGRAMS

Bloom promoted teaching to mastery before having students move on, and some mastery-based programs have achieved notable success even with heavily disadvantaged students. Perhaps foremost among them is the winner of the largest educational experiment ever run, Project Follow Through. Because many kids leaving Head Start preschool programs in the 1960s soon lost the gains they had made, the US Department of Education sponsored the project, which involved hundreds of thousands of elementary students across the nation. Parents chose one of the available teaching models for their school, and then a comparison school was selected in that area. Twelve different programs competed, and all students received the same thorough testing afterward. Developers of all programs agreed on the need for critical thinking—the ability to understand and apply higher-order concepts, not just memorize facts.[37]

The winning Direct Instruction program relies on carefully designed materials with steps that are neither too large nor too small, and multiple examples for better generalization. Its features include teacher modeling of skills; a great deal of active student participation; learning to mastery; constant adjustment to student success or failure; and consequences including immediate feedback, praise, and natural rewards. (Some sort of positive reinforcement was, of course, part of all the programs, but it was a focus in Direct Instruction, and a high positives-to-negatives ratio was built in.) The final evaluation showed that, compared to students in the comparison schools, Direct Instruction was the only program to produce substantially stronger gains in all three test areas: basic skills like spelling and simple math, higher-order cognitive skills like reading comprehension and math problem solving, and self-esteem (no surprise, since we know self-esteem depends on actual achievement). While Direct Instruction schools averaged large leaps in their reading scores in particular, most of the schools in the other programs scored significantly *worse* than their comparison schools. All four analyses of Project Follow Through's data have confirmed that Direct Instruction demonstrated the best outcomes. (It was not chosen for funding afterward, but continues to be available and successful.)[38]

Direct Instruction has been joined by other successful programs using effective methods. The Harlem Children's Zone's Promise Academy Charter Schools (elementary and middle) require extended hours and extra practice and use extra incentives, for example. Students significantly behind their grade levels on state tests catch up in a few years, particularly in math. Comparing students who lost the charter lottery to those who got in, Harvard economist Roland Fryer and a colleague recently found large gains for the Promise Academy kids: "The effects in elementary school are large enough to close the racial achievement gap in both mathematics and ELA [English Language Arts]."[39] Another example is the largest US charter-school network, the Knowledge Is Power Program (KIPP), founded by two men who adopted some of Rafe Esquith's techniques, namely, extended school hours, extra practice, and a point system (KIPP dollars). Like Promise Academy, KIPP schools have been documented to help prepare disadvantaged youngsters for college, often overcoming their late starts.[40]

MORE ON MOTIVATION

In one study, students took a test either for a grade or simply for the sake of learning. Is anyone surprised that they studied more and did far better with a grade for motivation?[41] Still, while end-of-term grades are supposed to motivate, they're inherently delayed consequences, and they don't work for everyone any more than studying is intrinsically rewarding for everyone. Teacher or parent approval and disapproval don't always work either. (These of course are also artificial consequences, like grades.)

Here is a controversial idea: regardless of the program implemented, motivate kids by paying them *real* money. To investigate, Roland Fryer ran randomized controlled trials of very different pay-for-performance ideas in New York; Washington, DC; Chicago; and Dallas. Thousands of kids and millions of dollars later, only the Dallas and Washington approaches showed significant *academic* success.[42] No wonder, Karen Pryor suggests: the unsuccessful approaches relied on delayed *outcomes* (not behaviors) like better scores on the end-of-year standardized tests that all the kids took—and (necessarily) long-delayed consequences. Behaviors in the successful approaches were clear-cut, and the rewards were far more immediate. In Dallas, second-grade kids simply read books and tried to pass online quizzes for a fast two bucks. As Pryor noted, test scores shot up: "It was as if the kids had had another half year of schooling. And it cost Dallas about $14 a kid."[43] Separately, education analyst Larry Cuban and cognitive scientist Daniel Willingham agreed, noting that the successful Washington approach also focused on actual behaviors like attendance and on-task behavior, with rewards within two weeks at the latest—far more quickly than the two unsuccessful approaches.[44] Leaders at KIPP schools have also found that more immediate rewards work better.[45]

B. F. Skinner favored (and constructed) carefully designed educational materials that provided frequent questioning, individual pacing, and the immediate intrinsic rewards of achievement and progress. He thought these natural rewards would usually be sufficient for normally developing children, and artificial rewards unnecessary, not even praise.[46] (Indeed, why add extra consequences if students are already learning efficiently?) Unfortunately, natural consequences alone seldom cut it in practice, and support for extra consequences crosses the spectrum: from experts like Roland Fryer, Albert Bandura, and Benjamin Bloom to the leading researchers of intrinsic motivation.[47] Consistent with Carol Dweck's research (see chapter 11), education expert Mark Morgan commented, "The evidence seems to support strongly the hypothesis that rewards that emphasize success or competence on a task enhance intrinsic motivation."[48] That's also the conclusion of the largest review of the research.[49]

CONSEQUENCES AT WORK

From Steven Levitt and Stephen Dubner's book *Freakonomics* we find this observation: "*Incentives are the cornerstone of modern life*. And understanding them—or, often, ferreting them out—is the key to solving just about any riddle."[50] They are certainly an important part of the working world.

We spend most of our time at work, and as a consequence we are able to keep food on the table and a roof over our heads. Lots of other consequences are found at work too—ideally some natural reinforcers like variety, a level of control (recall from chapter 1 the British government officials who benefited), and appreciation (recall from chapter 11 the lead printer who had to solicit praise because otherwise she didn't get any).[51]

Principles of consequences were applied long before the science existed, of course. Whaling crews worked for shares of the profits instead of the daily wages of other ships. The result: more risk-taking, which is quite important if you're trying to harpoon one of the largest creatures on the planet while bobbing about in a small boat on a big ocean.

Early twentieth-century ornithologist Robert Cushman Murphy observed that the captain of one old-time sailing ship, the whaler *Daisy*, added another unusual motivation system. He would call for a volunteer, and several sailors would immediately rush over. Easy tasks went to the first, dull ones to the last, and unpleasant ones to members of the watch who failed to volunteer. The crew quickly caught on.[52] (Know any bosses like this?) Employees can take unfair advantage, too: As novelist Jonathan Franzen found, US mail carriers who were conscientious and finished early used to get the leftover work of the less conscientious. Inefficient workers who finished late used to be rewarded with overtime pay![53] A staff member in one academic department observed that many junior faculty failed to get tenure, and was able to get away with doing little work for them.

It doesn't take much to turn that around, though: in one famous five-year study, almost any change increased productivity on a factory line, due in part to the attention and the fact that the workers knew their output was under scrutiny.[54]

Ideally, managers and employees alike have incentives to behave fairly—with positive expectations for all. The lessons of Rosenthal's self-fulfilling classrooms apply directly to the workplace. As consultant Aubrey Daniels put it, "the mission of a boss at any level of the organization is to 'create successful employees.'"[55] Here are a few examples illustrating how the science of consequences has helped.

That magic 5:1. Business bestsellers like *The One Minute Manager* are full of positive reinforcement: one minute praisings of behavior (with true feeling, of course), gradual shaping of good performance, and other consequence standards. One of the foundations: "Consequences maintain behavior."[56] Indeed.

If the positives-to-negatives ratio is important in marriage, parenting, and education, it's not surprising that it figures in the workplace, too. Back in 1950, Robert Bales's classic *Interaction Process Analysis* covered its importance in small groups like business teams.[57] Tom Rath and Donald Clifton concurred in *How Full Is Your Bucket?*,[58] and the ratio is critical in current mathematical models from business schools.[59] Sports, too; witness the Positive Coaching Alliance.[60] Again, it's not as easy as it sounds. Consultant Daniels, an undoubted expert at the office, still had to remind himself to be more positive when teaching his son to drive.[61]

Schedules. Would you make more sales calls if paid by the hour or on commission? More widgets, if paid by the hour or by the widget? In the field and at the lab, studies show that work-based ratio schedules get more out of people than equivalent time-based ones. Researchers have checked out office tasks like entering data, restaurant serving, factory work, even counseling.[62]

But widgety piecework—being paid by the piece (see chapter 5)—is easily abused, and was in fact abused historically. With such a strong incentive, workers would push themselves to earn more. Management could then increase the required number of pieces per pay unit, so that workers never gained from their productivity. They could end up working at an exhausting pace for a bare living, risking injury like the old-time whalers. Schedules are powerful.

Schedules themselves aren't inherently problematic, though; it's how they are used. Sweatshop workers could be—and are—exploited on an hourly wage system as well, a totally different schedule. On the positive side, in early British Australia, convicts at first worked long, set hours and did the minimum. Switching to a work-based ratio schedule instead—like piecework—turned out to be life changing for some. The convicts could finish their work quotas early if they were diligent, then work "overtime" for pay or switch to their own projects, such as raising produce they could sell.[63]

We're starting to take better advantage of schedules in business applications. For variable schedules to keep us working more steadily and happily, for example, researchers find that occasional easy rewards are helpful (see chapter 12). This knowledge has been applied in the design of educational software and video games, and in vigilance tasks like security screening and quality control. In a recent study simulating baggage screening, participants hit a space bar to produce an x-ray image of luggage. Including prohibited items on a variable schedule rewarded careful observing, producing significantly higher rates of work than a long series of baggage scans with no such items. It's simply harder to stay motivated if you never find anything. As the authors suggest, "management systems could systematically and frequently plant artificial signals for workers, set goals for signal detection performance, and reward high performance."[64]

Relatedly, some business authors have noted how hard it is to get through a long project, in essence a cumbersome fixed-ratio schedule. Reinforcing progress can help. We saw in chapter 7 that signals (such as progress markers) sometimes ease the way, but under other circumstances, they just make painfully clear how much farther there is to go, slowing things down. Using what we know about how signals and schedules work can help make work easier as well as more efficient—a win-win deal.

Extra incentives. What's an employer to offer? Creative incentive systems abound, and even little incentives can have big effects. Lotteries for cash prizes were offered for good factory attendance in one study. During the four-month program, absenteeism declined 18 percent, while it increased 14 percent in comparison groups at other factories.[65] A point system at a grocery store cut the shifts missed per week in half.[66]

Just as in the classroom, these sorts of systems can be positives-only, negatives-only, or combinations of positives and negatives. The programs need to be planned and evaluated carefully for fairness and acceptability, of course. They can be rollicking successes appreciated by most employees, or they can be dismal failures.

Indeed, while "Employee of the Month" programs may seem like obvious choices, consultant Aubrey Daniels discourages them. He explains that they offer only delayed and infrequent rewards, they are not based on specific behaviors, very few can win, and they can create rivalries instead of cooperation. What's more, not everyone finds the public notice reinforcing.[67] A recent study found that one Employee of the Month program had no measurable effect even when a financial prize was offered in addition to recognition.[68]

We can do better. In one study, package-loading teams competed in a baseball-style team competition, with standings posted publicly and free dinners provided for the winning teams. All three measures of accuracy improved: for example, the number of boxes per sorting mistake increased from about 4,800 to 5,600. Most of the workers enjoyed it.[69] Even here, though, experts like Daniels prefer more cooperative approaches, like competition against a fixed standard so there can be more winners and less rivalry.

At a food-distribution warehouse, workers packaged their products with credits for accuracy and speed, and subtractions for mistakes—a more positive version of a system that had been negatives-only at first. A progressive schedule was used up to a point, so the criteria got more difficult as the workers met the requirements. Errors decreased 10 percent, and employees appreciated being able to earn extra money: more than 90 percent wanted the program to continue. Did the company benefit, considering its extra costs? The net savings during the study period was almost $10,000.[70]

Safety. On a different scale, safety lapses at work cost lives, millions of injuries per year, and billions of dollars. Shipyards, paper mills, chemical companies, electricity-distribution centers, factories, you name it: extensive research has demonstrated the success of a variety of consequence-based safety programs.[71]

Rules alone may not work much better in a business than in a classroom—if consequences don't back them up. That goes for good safety practices, too, as one study with hospital workers demonstrated. There are right ways and wrong ways to transfer disabled patients. Adding simple feedback with praise brought a noticeable increase in safe transfers, and the workers liked the procedure enough to recommend it.[72]

When truck drivers are paid by the mile rather than the hour, they are rewarded for pushing themselves, driving too long and too fast (like piecework, a similar "ratio"/work-based schedule). According to one study, they were about twice as likely "to doze or fall asleep at the wheel" as hourly drivers.[73] Union drivers are usually paid by the hour rather than by the mile, but nonunion drivers are subject to what's most economical for employers. (Consequences at work.) At the time of another driving study, pizza deliverers had much higher crash rates than the general public. Part of the problem was the ratio schedule of reinforcement: the more pizzas delivered, the more tips and commissions, so risky driving got rewarded. Simple interventions like feedback about their driving helped, and the rest of us could breathe a little easier.[74]

Finally, one of the most dangerous jobs in the world is mining. In a project lasting over a decade, three researchers set up a positives-and-negatives point system at two open-pit mines (uranium and coal), each with hundreds of employees. Workers earned trading stamps for using safe practices and avoiding injury, suggesting new safety rules, and belonging to groups who avoided injury. The group rewards meant that workers had an incentive to help each other follow safe practices. (After all, these practices do slow things down.) Injuries meant a loss of stamps, as did a failure to report accidents. The stamps could be exchanged for a large selection of items.

The number of days lost from work plummeted at both mines after the system began—to only 11 percent of the original level at one mine and an astonishing 2 percent at the other, both well below national mining levels. The costs of accidents and injuries dropped by about 90 percent, and the savings were so large that they covered the program costs many times over.[75] Most important, the program may well have saved lives.

Economics and marketing. Switching to the demand side, "behavioral economics" combines economics with the science of consequences to understand the choices we consumers make. That means understanding reward value. As an example, we have already seen mathematical descriptions of how value drops when consequences are delayed (see chapter 12).

Changes in reward value can be simple—don't go grocery shopping when you're hungry—or complex. Take economic "elasticity," the degree to which a consequence is rigidly demanded or has substitutes. You might not care if you drink tea, coffee, or a cola for breakfast (elastic), but you need your caffeine (not elastic). A lot clearly depends on what other choices are available. And a lot depends on rules. If you care about environmental issues, you might pay more for shade-grown coffee, or, switching to lunch, an organic, locally grown tomato. Rules, signals, schedules, emotional associations . . . sometimes the simple "naked" reward value of a consequence seems the least part of it. The possibilities are dizzying.

Some of us respond most strongly to a change in price or effort, others to how often the rewards come, and still others to need-it-now impatience with delays. In economic terms, our "demand curves" are different. Given different choices, completely different factors can loom large, then change the next day. Marketing experts face the challenge of creating the greatest reward value for the greatest number, putting the good in "goods"—or at least making us think so.

Brands are one of their successes. Given blind taste tests, for example, hordes of volunteers have been astonished to discover that they rejected their favorite brand of cola.[76] Brands develop a life of their own—their own emotional associations and reward value—through actual experience, social modeling, rules, and, of course, ads. People feel so strongly about them that they even wear T-shirts proudly displaying a favorite brand name, providing free advertising.

In order to construct effective ads that enhance reward value, marketers take advantage of every inch of accumulated scientific knowledge. To create classically conditioned emotional associations, for example, executives have paired products with celebrities for over a century. (And no wonder, since studies confirm that this strategy works.[77]) When nineteenth-century theater star Lillie Langtry endorsed Pears' Soap, she may have been the first female celebrity to appear in an ad.[78] Similarly, ads using positive-emotion words like "new," "easy," and "amazing" sell better.[79] Tug on heartstrings with direct emotional appeals: show a child in pain and your product gets associated with ending the pain. Show a happy couple for happy associations. (Fear can be even more powerful, as creators of political ads know.) Beautiful scenery, attractive models, and catchy tunes are all positives. Motion gets our attention (the Pavlovian orienting response) and then rewards our continued, fascinated gaze. Pulling out all the stops, the US Super Bowl ads may attract as many fans as the football.

Reasoning has brought us science fiction turned reality, and yet our reasoning "blind spots" still trip us up. Marketers know all about them too. Eighty-five percent lean hamburger meat is the same as 15 percent fat, but the gut impact of these labels is different. The way positives and negatives add up isn't always straightforward either. Language framing can play tricks: sellers may profit more through offering rebates (wow, a free lunch!) than through simply setting a lower price, for example. On the other hand, having to wait a long time for the rebate can turn that equation on its head. Because of delays, even how we pay makes a difference: Using a credit card emphasizes the vast distance to actually having to pony up real money. Shelling out cold, hard cash instead can nix the deal.[80]

Finally, the ubiquitous point systems again: marketers would hardly overlook them. Some of us remember the old S&H Green Stamps program (now S&H greenpoints) in which accumulated books of stamps could be exchanged for different items. (I know that dates me.) These days, frequent-flyer programs may be the best known of the many extra incentive programs that reward customer loyalty. Credit-card rewards programs have also become popular and are sometimes linked to frequent flyer programs. There are still plenty of proof-of-purchase plans, too. The current nonprofit Box Tops for Education program, for example, has provided millions of dollars for US schools by partnering with manufacturers of grocery goods.

An intriguing schedule effect has been documented for one such system: to "win" a free coffee, buy a number of coffees and get a card checked off. The question: Is it better to provide a card where twelve purchased coffees need to be checked off, with two already docketed (freebies!), or to provide a blank card requiring ten coffee purchases? They are exactly the same, of course, but this is a fixed-ratio schedule. That means having a start makes it seem easier—you have already begun the "run" phase of the characteristic "break-and-run" pattern (see chapter 5). And that's what the research showed, hands down.[81] What fun.

Rewards at work: the flip side. Marketing dangerous products . . . Poorly planned incentive programs that boomerang . . . Any science can be woefully neglected, abused, or misused. But what could be wrong with praise on the job? Surely anything that lightens the load is welcome?

In his bestselling exposé *Fast Food Nation*, Eric Schlosser noted that many fast-food restaurants hired mainly part-timers so they didn't have to pay benefits (consequences at work). These employees are paid minimum or near-minimum wage, and they often move on when they can. To compensate, at least one large chain taught "stroking"—deliberate praise that made these low-paid workers feel valued.[82] It can work for a while, helping to keep them on and avoid training costs for replacements. If the praise is sincere, it can work longer.

Employees regularly choose lower-paid jobs, after all, if they are more rewarding in other ways: child care, for example, or zoo jobs for animal lovers, or nonprofit work for activists. My Peace Corps experience showed me plenty of inspiring examples. Indeed, many people who can afford it do full-time volunteer work for no pay whatsoever. Employees with fewer choices, however, take what paid jobs they can get and make the best of them.

Minimum wage, with or without praise . . . Both have their down sides, but which would you choose?

Chapter 15

HELP FOR ADDICTION, AUTISM, AND OTHER CONDITIONS

I am now the most miserable man living. If what I feel were equally distributed to the whole human family, there would not be one cheerful face on the earth. Whether I shall ever be better I can not tell; I awfully forebode I shall not. To remain as I am is impossible; I must die or be better, it appears to me.
     —Abraham Lincoln, in a letter dated January 23, 1841

Depression is one of the many psychological problems that consequences can help cause—and help relieve. I will discuss a number of such conditions. In each case, let me make clear that, following the systems approach adopted in this book, many factors are involved in their causation. I will be emphasizing the ways in which the science of consequences adds to the treatment repertoire.

CHURCHILL'S "BLACK DOG": DEPRESSION

Abraham Lincoln, Virginia Woolf, Winston Churchill . . . depression has brought misery to the famous as well as to everyone else, throughout the world, throughout history. Hippocrates (of medicine's Hippocratic Oath) sadly described depression back in ancient Greece. One of its defining features is devalued consequences: no reason to live, nothing to look forward to.

Rewards lose their savor. Negatives overwhelm. The positives-to-negatives ratio upends.

And for once, our proudest skill hurts more than helps: thanks to language, we can enjoy *endless* worries and regrets. Anticipating trouble or miserably ruminating about the past multiplies our suffering many times. We can even foresee our own deaths.

Vicious cycles. Short of death, lose the people you love, or your job, or your dearest hopes, and you not only lose many of the positives in your life but you're also more likely to suffer from depression. The same thing can happen if you begin life weighted with disadvantages—enduring abuse and poverty, for example.[1] Or just get buried by too many small- to medium-sized negatives.

Once you're down, spreading activation (see chapter 10) brings yet more negatives to mind, one black association after another. Here again, language can hurt: it expands the range of negatives. As a patient mistreated in a hospital for the poor, for example, novelist George Orwell was struck by an emotional memory for Tennyson's tragic poem, "In the Children's Hospital." Though Orwell hadn't thought of it for many years, the story and even many of the lines resurfaced, prompted by the associations.[2] Similarly, cancer patients were once thought by some to have been distinguished by unhappy childhoods (the "cancer personality").[3] Instead, anyone in pain and unhappy is simply more likely to recall unhappy things.[4] This can be helpful to keep in mind.

It's also helpful to recall that, like beauty, what's negative is in the eye of the beholder. In one study, annoying highway noise was no problem for those nearby homeowners who had positive associations: handier stores, convenient services, and more jobs.[5] Shakespeare was on to something when he had Hamlet say, "There's nothing either good or bad but thinking makes it so." What's more, associations and perceptions can change in an instant: You think your parents were happy together, then you discover the shocking truth. Suddenly your whole life looks different.

Tell someone. Talk it out.

Talking it out. It helps simply to see that the world doesn't come to an end when your pain is in the open. In fact, supportive ears are lifesavers. A much-cited *Science* study found loneliness to be riskier than smoking, and an increased chance of depression has been shown to be one of the reasons.[6]

Besides bringing rewards in itself, talking helps Pavlovian processes soothe churning emotions: it associates painful topics with less painful feelings to weaken the impact. New ways of interpreting what happened mean new emotional associations. Accepting negatives defuses them. From one man who told a therapist about his struggles over losing a parent: "Once I put words to it, it was as if it didn't have power over me. Whereas before, it had been an undifferentiated mass of anxiety, sadness, and fear."[7]

If you lack someone to talk to, innovate. Keeping a journal can help, even writing a letter that's never sent.[8] In Guatemala, children traditionally told their troubles to small, colorful "worry dolls," put them underneath their pillows, and slept the troubles away.

Learned helplessness. While worries are always going to be with us, we have to accept negatives that we don't think we can change, or at least learn to cope with them. But sometimes there's a price to be paid.

Clara Barton helped lead the Union nursing effort in the Civil War and then fell into a five-year depression. She was prevented by barriers of sexism from accomplishing anything further. When she finally had the opportunity to be useful again, her depression ended, and she went on to found the American Red Cross.[9]

Surely we can all relate, if only in smaller ways. After I developed knee problems, nothing I tried helped much. Frustrated, I simply gave up for several years—no rewards to keep me going. How discouraging (and unwise). In the lab, self-esteem researcher Carol Dweck and a colleague showed that when children had no luck with math problems that were unsolvable (unbeknownst to them), they generalized to a set of problems that *could* be solved—and didn't even try.[10] Similarly, when college students could not escape a loud noise, they simply endured it, while new arrivals took the control that had become available.[11] "Learned helplessness" is so widespread that under comparable circumstances in the lab, fish develop it, even roaches.[12]

Fortunately, it's possible to "inoculate" against such helplessness to some extent.[13] People and animals alike, when we learn we can persevere to escape bad situations, we are less likely to suffer in silence. Experience with variable schedules has been shown to help; after all, persistence in the face of disappointment is their hallmark.[14] (Problematic for gamblers. Beneficial here, where persistence is critical.)

Confined for the duration, we assert ourselves by taking what control we can. Just being in charge of a potted plant helped nursing-home residents stay healthier (see chapter 1). Similarly, if you are suffering from depression and at your wit's end, simply making an appointment with a therapist brings some relief.[15]

Bringing back the sun. Enduring rough times, I once glanced up to discover sunlight filtering beautifully through the trees. Other positive associations started fluttering their wings, and my mood lifted for a while. It may seem obvious, but research shows that people who have recovered from depression are likely to relapse unless they get reinforcers back in their lives somehow. Do we need a minimum number? Or, perhaps, a "magic" positives-to-negatives ratio? We've seen that ratio everywhere else, after all. (We've also seen the neurophysiological benefits of reinforcement. Consequences, therapy, antidepressant drugs, they've all been shown to change the brain.)

Famous biographer and depression sufferer James Boswell once wrote: "Did I not imagine myself doomed to unceasing melancholy? . . . And yet, my friend, I am now as sound and as happy as a mortal can be. How comes this? Merely because I have had more exercise and variety of conversation . . ." (letter written July 23, 1764).[16] But can the benefits last?

While most depression therapies try for more contact with rewarding activities, it's the primary focus of "behavioral activation therapy." The idea is simple: boost your positives-to-negatives ratio, getting reinforcers back in your life and the benefits of their positive associations. (The therapist also recommends methods to deal with insomnia, loss of appetite, and other common depression symptoms.) One meta-analysis covered randomized controlled studies of this therapy, over 2,000 patients in all. It found that it could be considered "well established," the highest level of evidence basis.[17]

This therapy can help severe depression as well as milder cases. For example, clinical-psychology researcher Carl Lejuez and his colleagues worked with twenty-five severely depressed patients in a hospital, rewarding increased activity: simple things like reading, chatting, going for a walk, and cleaning their rooms. They improved significantly.[18]

Indeed, exercise alone can be a reinforcer with therapeutic benefits all on its own. In one randomized controlled study with depressed people over the age of fifty, it also produced significant improvement[19] (although reviews of the entire exercise-depression literature have found inconsistent support[20]). Bring back the sun and keep it shining.

ANXIETY AND FEAR

Darkness descends when we anticipate the worst. Among the most powerful consequences are the things we fear: we work very hard indeed to avoid them. But we can't dodge the perilous future.

Few of us are able to resist harboring some fears, often with good reason. While we learn to cope with most of them, even garden-variety fears can be limiting. Social fears mean fewer friends; fear of flying means seldom seeing your favorite cousin. And some fears can become incapacitating.

Specific phobias—spiders, heights, enclosed spaces—are the most common type of anxiety problem. Fortunately, the standard fix is very effective: facing what you fear. (No one said it was *easy*, just successful.) Therapeutically, this is normally done in carefully measured steps, rather like shaping. By deflating its Pavlovian emotional power, this approach also changes the feared thing's consequence value. "Exposure"-based therapy has been the treatment of choice for years, confirmed by a recent meta-analysis of randomized controlled studies.[21]

Exposure doesn't have to be direct. As we've seen, thinking about something can have effects similar to experiencing it, physiological effects included (see chapter 8). Ironically enough, that gives us another way to vanquish fears. If you're afraid of heights, imagining yourself standing on a cliff edge will upset you. However, force yourself to do so until you are calmer, using deep breathing and other relaxation techniques, and you've taken a step toward the real cliff edge. These days technology like "virtual reality" lends a helping hand. Even so, if an arachnophobe touches a toy tarantula while working through a virtual-reality spider sequence, the therapy works better.[22]

Blood-needle phobias are relatively common, and people who suffer from them, like journalist Letty Cottin Pogrebin, understandably tend to avoid medical testing. Pogrebin's kids got their first aid from others, not their mother. Going into shock because of her phobia finally got Pogrebin heading for help. After seven years of other therapies, she found a therapist who used exposure methods and was cured in ten weeks[23] (and that's on the long side by current standards).

As with learned helplessness, we can immunize ourselves against potential fears. Get experience with whatever it is, pair it with positives, watch other people happily doing it, read about how safe it is. These work for animals, too (well, not the reading part; yes, animals get phobias).

Children sometimes acquire their parents' fears, but parents can help ensure they don't by taking advantage of such techniques. For example, researchers Susan Mineka and Richard Zinbarg described two cases of teen-aged girls attacked by dogs. Only the teen who had had little experience with dogs developed a phobia. The other girl did not, being protected by her history of good experiences just as laboratory research predicts. Similarly, given a painful, upsetting encounter at the dentist, kids who had had a number of pleasanter experiences were less likely to develop phobias than children new to the tooth-care scene.[24]

GETTING UNHOOKED: ADDICTION

The science of consequences helps researchers study the ups and downs of the reinforcing value of drugs—and predict which ones are likely to be abused.[25] Unexpectedly, they discovered that less-than-obvious events can change a drug's reward value. For example, being defeated in a fight can raise the reinforcing power of cocaine—for rats.[26] (Yes, animals will give themselves cocaine.) How about for people? It would be useful to know, and researchers are working on it.

Pivotal fact: Addictive substances can be so powerfully reinforcing that addicts may steal, even kill, to get them.

Smoking. Cigarette addicts don't kill—their substance of choice is legal—but smoking kills hundreds of thousands in the United States each year.[27] Why would anyone start? Many influences are obvious: As a kid, if your friends smoke, the social rewards (and pressures) are large. Less obviously, researchers found recently that three- to eight-year-olds whose parents smoke are substantially more likely to prefer the smell of cigarettes over a neutral odor, compared with nonsmokers' kids.[28]

At any age, rules alone frequently fail to guide us, however well supported, however much we know they should. Often, the dreaded consequences they describe are long-delayed, and that's part of the problem. Because the written warnings on cigarette packs are a good example, dozens of countries now require, in addition, pictures of diseased lungs and other emotionally striking images.[29]

Most smokers eventually get the message and quit for a while. Staying off is what's hard. If a person has a bad day, on top of dealing with nicotine withdrawal, it's tough not to reach for that easy reward, a quick smoke. (Who among us can't relate?) How to help would-be quitters hang on for the long term? In Australia, no lung transplants are available to people who smoke or have any other substance abuse. Less drastic, behavioral pharmacologist Jesse Dallery and his colleagues devised a system based on "deposit contracting," similar to Ian Ayres's stickK.com (see chapter 12). Smokers who want to quit put up their own money and earn it back if they stay off.[30] Relatedly, researcher Kevin Volpp and his colleagues recently published a successful randomized controlled study of cash incentives for quitting in the *New England Journal of Medicine.*[31]

In another recent randomized controlled study, "hard-to-treat" smokers were the focus. These addicts had failed to kick the habit during their baseline visits to a clinic, even though they had volunteered for a smoking cessation experiment. (A breathalyzer determined whether they had smoked recently, to back up their own reports.) Half of them had tried to quit at least twice before. For the next sixty visits, smokers in one group were given cash for entirely avoiding smoking on the previous day—a "cold turkey" type of approach. Smokers in a "shaping" group got cash for gradually cutting back. Nearly half of the hard-to-treats responded well to the shaping approach, but only a quarter to going "cold turkey."[32] Along these lines, in another random-assignment study, fixed versus progressive schedules of reinforcement were compared. Participants on either schedule were almost twice as likely to quit as those in the control group. But participants on the progressive schedules (which gradually get leaner) were the least likely to relapse after the program ended—by a wide margin.[33]

Cues for smoking become rewarding in themselves in the same ways we've seen before. They also increase the immediate reward value of cigarettes by producing cravings (as confirmed in lab studies):[34] See your pals smoking and get the urge. Visit your favorite bar and get the urge. That makes it even harder to stay off once you've quit, because cues are everywhere. Trying to avoid them can turn quitters into hermits.

Prevention is of course the best solution, and that means discouraging kids from picking up the habit. The Good Behavior Game, which rewards groups of kids for appropriate classroom behavior (see chapter 14), is one of a number of childhood programs that succeed in doing just that. These programs help head off illegal drug abuse, too. (And on a larger scale, societies with less poverty and more equality of opportunity also have less illegal drug abuse.[35])

Illegal drugs. Cutting the demand for illegal drugs—that is, their reward value—would have huge benefits for all of us. The overall financial burden alone is breathtaking, estimated at almost $200 billion per year just in the United States.[36] But once addiction takes hold, most treatments fail most of the time. Why?

Classical conditioning cues apply to cocaine, meth, and other illegal drugs, as well as legal ones. Take old haunts and habits, add the drug's huge reinforcing power, plus the negative of the nasty withdrawal symptoms. The result is a formidable challenge even for addicts who strongly want to quit. Drugs may be the most rewarding thing in their lives.

For twenty years, one of the more successful treatments has provided healthier alternatives. As introduced in chapter 12, to earn vouchers, addicts provide drug-free urine samples. The vouchers can be exchanged for specific items. (Some versions add employment opportunities.) For extra motivation, the longer addicts go without a drug-positive sample, the larger the voucher they earn. Relapsing means falling back to the initial smaller amount. Meanwhile, the natural rewards of being drug-free are also (we hope) piling up. Not surprisingly, bigger vouchers bring bigger success rates.[37]

A recent meta-analysis of randomized controlled trials of treatments for illegal drug abuse concluded that this consequence-based approach ("contingency management") had "the strongest effect."[38] (And England's National Health Service recently endorsed it.) While these programs are cost-effective in the long run, though, they can be costly in the short term. To make them more affordable, lottery reward systems are being tried—staying drug-free earns the chance to win a prize.[39]

AUTISM

Roughly one of every 100 children born in the United States is now diagnosed with an autism spectrum disorder. To varying degrees, these children show delayed language, poor social skills, and repeated behaviors like hand flapping. Apart from a high-functioning group (Asperger's syndrome), most do poorly on intelligence tests.

For years, most autistic people were institutionalized, often unable to communicate or handle everyday living skills. That's all changed because of a revolution brought about by the science of consequences: many can now live and work in their communities (with some assistance), leading vastly improved lives, enjoying some degree of autonomy. And a surprising number of children diagnosed with autism have made even greater gains.

The foundational study came out twenty-five years ago. After decades of prior development, psychologist Ivar Lovaas and his colleagues provided nineteen young children diagnosed with autism with one-on-one teaching for forty hours a week (now known as Early Intensive Behavioral Treatment). After two years, at six to seven years of age, IQ in this group of children had increased by 20 points on average, to close to normal (85 is considered normal). Nine of these children gained around 30 points, achieved scores over 85, and were mainstreamed into regular classrooms. Similar children in an untreated group and a less-intense treatment group did not improve, with average IQs staying in the 50s. At eleven to fourteen years of age, eight of the successful group of nine were still going strong, still mainstreamed and testing within the normal range (IQ and other behaviors).[40]

This was not a fluke: other studies have achieved similar results, sometimes with older children. Hundreds of studies of early intensive behavioral treatments have been published. In a recent review, only these treatments met the standards to be considered "well established," with randomized controlled studies and other experimental evidence in support.[41] (A related approach, Pivotal Response Therapy, met the criteria for "probably efficacious.") Accordingly, consequence-based methods have been described as the "treatment of choice" by the US Surgeon General and endorsed by many professional organizations such as the American Academy of Pediatrics.[42]

In teaching language and social skills, all the methods of chapter 14 are used—shaping and modeling, for example—and then some. Nothing can be taken for granted: imitation may have to be taught, for example. In one study, four autistic children learned to imitate vocalizations, toy-play movements, and "pantomime" movements on a variable reinforcement schedule—enabling them better to maintain and generalize this critical skill.[43]

What about language? Catherine Maurice, mother of two autistic children and author of *Let Me Hear Your Voice*, describes getting her son to communicate: "I can hold these coins, and not give them to him until he looks at me and makes some sound for each one. The basic technique is always to get him interested in something . . . and then to use that something to keep pushing the idea of language, communication. . . . I say something, and when I do, that bothersome mommy gives me what I want."[44] And on through an immense number of steps to skills like handling sophisticated conversation.

Teaching is planned and incidental, at home and in the classroom. Positive reinforcement is always the focus. If the program is not working, change the setting or the schedule, try using smaller shaping steps, more (or less) prompting (partial cues), or more (or less) praise, make the task easier or more challenging, or break it into smaller parts . . . and so on. Adapt. Don't give up.

Definitely don't give up with a previously intractable problem. Some autistic people hurt themselves, most commonly knocking their heads against things, hitting their heads with their arms, or biting themselves (a terrifying thing to see). One little girl "had hit herself so forcefully and so constantly that she had caused frontal-lobe brain damage, punctured her eardrums, and damaged her eyes so badly that she was legally blind."[45]

Scientists have met even this challenge, relying on functional analysis to determine *why* self-injury happens—and it turns out to be consequences. From a review of 152 self-injury studies, about a quarter of the cases were due primarily to the intrinsic reinforcement of the activity (odd as that may seem). The rest were due mainly or in part to the rewarding attention the self-injury brought. (Not an isolated phenomenon: recall that non-autistic kids sometimes misbehave in order to get attention.) In many of the cases, escaping from an educational or living-skill task was an important consequence.[46] Functional Behavior Assessments for such self-injury cases are now required by US federal law.

Knowing why self-hurting is occurring lets the treatment be correctly targeted. If a child hits himself because that gets him out of language practice, make practice more rewarding. If it's primarily to get attention, teach better ways of getting attention ("Thanks for raising your hand!"). If it's intrinsically rewarding, reward healthy alternatives more. Sometimes the self-abuse can be directly prevented: wearing gloves stopped a girl's face scratching while other behaviors were reinforced, then the gloves were removed for gradually longer periods. It worked.[47] Indeed, a review of almost 400 self-injury studies concluded that these consequence-based approaches are highly effective.[48]

Finally, autistic children tend to have trouble with "joint attention," a skill that's critical for language learning. Joint attention means what it sounds like: sharing attention to an event with someone else through gestures, eye-gaze signals, or comments. It sounds straightforward enough to do: normal kids, seeing something exciting like a low-flying airplane, point to it and look at nearby adults to call their attention to it. This behavior sequence is rewarded by the attention as well as the exciting event. Where in the sequence are autistic children missing out? *Starting* a joint attention partnership has been shown to be particularly problematic for them. Several studies, for example, have found that these children look at moving toys such as robots when they are activated, but don't share the excitement. Adult attention, it turned out, wasn't rewarding.[49]

Several teams of researchers have recently succeeded in teaching autistic children to initiate as well as participate in joint attention. When adult attention wasn't rewarding, that was also successfully taught. Nothing can be taken for granted.[50]

Careful, targeted teaching can accomplish wonders. Maurice describes what happened when a previously closed-off child "got it": "Suddenly, unbelievably, she stretched out her hand, pointed to the water, and turned around to look at me! I was thrilled."[51] After intensive consequence-based therapy, both of Maurice's autistic children were able to be mainstreamed.

ATTENTION DEFICIT HYPERACTIVITY DISORDER: DRUGS OR CONSEQUENCES?

According to the Centers for Disease Control and Prevention, nearly one in *ten* US children have been diagnosed with attention deficit hyperactivity disorder (ADHD).[52] It's a far more common problem than autism.

For sufficiently reinforcing activities, kids diagnosed with ADHD show no attention deficit: they tend to do just fine with rich schedules of immediate reinforcement. Lean schedules and delayed consequences are a different matter.[53] Not surprisingly then, ADHD children tend to be impulsive, valuing immediate over delayed rewards more than other kids.[54] Research on choice shows that these children are less sensitive to competing reward schedules over time (see chapter 7).[55]

Methylphenidate (Ritalin and related forms) and amphetamines (Adderall and related forms), stimulant drugs frequently used to treat ADHD symptoms, are regulated drugs of potential abuse. Related to ecstasy, meth, and cocaine, they share some common effects on the brain. They're better than caffeine at boosting test scores, and in one survey of almost 2,000 college students, fully one-third had used illegal ADHD stimulants.[56] These drugs can also be inhaled or injected recreationally, and there's a street market for them.

Amphetamines related to the ones prescribed for ADHD used to help long-distance truckers stay alert, and countless military personnel on both sides took them during World War II ("go pills"). ADHD drugs are less problematic than these particular amphetamines. Nonetheless, both types of ADHD drugs (methylphenidate and amphetamines) are banned in US commercial aviation, at the Olympics, and in some nations.[57]

There is an alternative: behavioral treatments. Almost forty years ago, for example, three researchers compared methylphenidate to a consequence-based program that taught self-control. Three hyperactive children between eight and ten years of age went on a point system (chapter 14), with reading and math as the focus. For all three children, when the medication was stopped, hyperactivity increased (up from 20 percent to 80 percent during time samples). On the consequence-based program, hyperactivity dropped back down to 20 percent. In addition, math and reading scores both improved markedly, from a 12 percent correct baseline to 85 percent correct, far better than under the drug.[58] To bring lasting benefits, though, behavioral treatments sometimes take substantial amounts of time, and some sort of maintenance program can be required.

Over a decade ago, the National Institute of Mental Health sponsored a large trial of ADHD medications versus a consequence-based behavioral treatment. While all groups improved, the medication-alone and combined behavioral-treatment/medication groups did best. On the face of it, then, medication did better than behavioral treatment. The study is not simple to interpret, though.

To begin with, follow-ups of these children years later showed that while drugs produced short-term benefits, they did not bring any enduring changes. Overall, there appeared to be no differences between children who had taken the drugs and those who hadn't.[59]

Then there are the details of the study itself. William Pelham, one of the study authors, noted that the medication-alone group had its dosage increased by 20 percent on average over the fourteen-month duration, because children were not benefiting. They received medication throughout that entire period. The combined behavioral-treatment/medication group also took medication throughout, but did *not* have the dose increased. Their results were similar to or better than the medication-alone group—leading to the inference that the behavior therapy was beneficial even though it was given for a much shorter time.[60]

Raising another issue, children in the behavioral-treatment-alone group were not tested for improvement until four to six months after the main program had ended—hardly an equivalent comparison. The logic was that this level of behavioral treatment would normally be all that was available, whereas kids on drugs are intended to stay on them for years. (They stay on drugs because if they stop, they usually revert to the ADHD behavior patterns.)

Parents in the behavioral-treatment-alone group received training in positive reinforcement and timeout but, not surprisingly, varied widely in what they did. A point system at school was an important part of the main behavioral program at the start of the study, but it was discontinued long before the end. Despite these disadvantages, the initial behavioral-treatment-alone results were statistically indistinguishable from medication-alone in all but three of the nineteen measures of improvement.[61] No wonder questions have been raised about what conclusions are justified.

More recently, Pelham and his colleagues ran a meta-analysis of 114 behavioral-treatment studies and found striking benefits. (Unlike previous reviewers, they included a much larger group of experimental studies and stuck to consequence-based methods.) Considering all the evidence, Pelham accordingly recommended that behavioral treatment always be provided, with short-term medication use only, and he's been joined by other experts.[62] But controversy remains, and meanwhile millions of children are continuing to receive drugs alone.

BRIEF NOTES

Psychopathy. Everyone has heard about the shocking murders committed by psychopaths like Jeffrey Dahmer and Ted Bundy. While most people labeled as psychopaths don't break the law, let alone go to these extremes, still, they are far more likely than the rest of us to become criminals (particularly violent criminals). How the condition develops remains a mystery, but researchers are making progress.

One possibility relies on classical conditioning and its effects on consequences. Given a fear-inspiring experience, for example, classical conditioning normally occurs to associated signals. Not for psychopaths. It's as if thunder frightens them, but not the lightning that heralds it; shock, but not the warning buzzer that precedes the shock. Likewise, show them a bloody knife and they don't flinch like most of us: they just don't react much to emotional words or images.[63]

Might children with fewer fears have a harder time learning to avoid negatives, including breaking inconvenient rules? It's possible (besides shedding a whole new light on the benefits of fear).

What's more, human empathy means picking up on emotional cues in others and feeling what they feel (classical conditioning again). No wonder psychopaths don't have it. People who can't *feel* for others would understandably find it harder to *care* for others. These in fact are defining features of the condition, along with characteristics like a lack of guilt (another missing emotion) and a lot of lying (no fear or guilt to stop them).[64]

Does the missing classical conditioning contribute directly to psychopathy, or is it just a side effect? To help answer this question, psychologist Adrian Raine and his colleagues studied classical conditioning in young children, following them from age three to age eight, and then checking in again at age twenty-three. When kids repeatedly hear tones and then a loud, unpleasant sound, classical conditioning of the skin's sweating response normally occurs: that is, the signal tones alone should start to cause sweating. This form of learning worked well for most of the children, but not all. Those boys and girls with poor classical conditioning were substantially more likely to be aggressive at age eight— and to turn criminal by age twenty-three.[65] Further research is proceeding.

If this finding about classical conditioning holds up, that still says nothing concerning how it comes about—and we've seen what a tangle of factors contribute to our development. More hopefully, we have also seen the many benefits of an early enriched, supportive environment, and that's what's recommended as a possible preventive by experts like Raine.

Problems of the elderly. People suffering from dementia sometimes get aggressive, and it's hard to reason with them. What should we do? Consequence-based approaches can help. In a typical study, two nursing-home residents with dementia hit staff who tried to bathe or shave them: in a sad scene, three aides had to hold a protesting man's head while a fourth shaved him. When nurses encouraged and rewarded cooperation through simple praise, these experiences soon became far more pleasant for everyone, and one aide could handle the shaving.[66]

A recent review of 162 studies of dementia found that consequence-based "behavioral management" techniques were among only a few successful approaches for alleviating problems like aggression, bizarre behavior, agitation, and depression.[67] Under current law, behavioral methods are supposed to take precedence over the still-widespread practice of "chemical restraint"—drugs.[68]

Then there are unusual situations like the true case of Ed, a chronically ill patient who kept willfully ignoring medical advice and ending up back at the hospital. What was going on? As it turned out, his well-intentioned family viewed Ed as a burden and failed to conceal the fact sufficiently well. At the hospital, in contrast, caregivers did their utmost to help and cheer him. No wonder he did what he needed to do to get there.[69]

Another problem common in old age is memory loss. However, if the elderly so frequently have enough brain flexibility to recover from stroke (at least partially, see chapter 4), maybe there's hope for hanging onto our most prized possessions, our memories. The research consensus suggests that making exercise more rewarding is one way to help. Another approach is a type of memory training that minimizes errors, thus maximizing reinforcement rate.

A recent review noted that so-called "errorless learning" was first investigated as part of the science of consequences—based directly on B. F. Skinner's work with elementary schools, in fact. Learning lists, associations, and new knowledge keeps the brain active and seems to have general benefits. Using the errorless approach, helpful prompts are gradually faded while small, easy steps move learning forward, and any errors that do occur are immediately and gently corrected. The positives-to-negatives ratio is quite high, naturally. So far, the most benefits of this approach appear to come to people with severe memory problems due to brain injury, but it's an ongoing area of research.[70]

Children who won't eat. Another ongoing area of research focuses on kids who won't eat, a problem that can occur particularly if children are developmentally delayed or have certain medical conditions. It can't be fun to have a feeding tube stuck through your nose and down your throat. Nor is it fun for parents to have to learn how to tube-feed their kids. But that's the main alternative.

Several research reviews have established that consequence-based methods are the treatment of choice. It's a matter of functional analysis again, understanding what's happening and why, and taking advantage of all the positive consequences available. The basics seem like simple common sense but are not as easy to do as they sound. When a child eats normally, reinforce that. Ignore problem behaviors, which can be rewarded by attention. Gradually fade in foods that kids don't like (keeping the positives-to-negatives ratio high). Start with small amounts and build up. The good news is that success rates are impressive, and many kids can finally toss the tubes.[71]

Tourette's syndrome. In a final example, people with this condition suffer "tics": inappropriate words or actions, like a jerk of the head during a conversation or involuntary speech. (Relatively few swear uncontrollably, although that symptom is probably the best known.) Early studies suggested that consequence-based methods could help.

Tourette's sufferers can often tell when a tic is coming on, and it's not a good feeling. In one analysis, giving in to the tic gets reinforced because it ends that unpleasant signal. A recent randomized controlled trial from the *Journal of the American Medical Association* included over 100 nine- to seventeen-year-old children. The treatment was learning to do something incompatible once the signal reared its ugly head, thus averting the tic while still ending the signal. Avoiding situations that led to the signals was also part of the ten-week therapy, as was relaxation training. Parents were involved at all stages, praising youngsters for using these strategies. The significant benefits were comparable to those from medication, and nearly all the children who benefited were still doing well six months later.[72]

Sometimes knowing how consequences work can set us free.

Chapter 16

CONSEQUENCES ON A GRAND SCALE: SOCIETY, THE LONG TERM, AND THE PLANET

A. Men, women, and children.

Q. And babies?

A. And babies. And so we started shooting them . . .

Q. Why did you do it?

A. Why did I do it? Because I felt like I was ordered to do it, and it seemed like that, at the time I felt like I was doing the right thing, because, like I said, I lost buddies. . . . So, after I done it, I felt good, but later on that day, it was getting to me.

—Mike Wallace of CBS News,
interviewing a US soldier who participated in the 1968
My Lai massacre of about 400 unarmed Vietnamese people

Seeing what was happening, an army helicopter pilot and his two crew members landed in front of the Vietnamese and stopped part of the massacre. Thirty years later they were officially honored, and General Michael Ackerman declared that they "set the standard for all soldiers to follow."[1]

OBEDIENCE AND DISOBEDIENCE

Bureaucrats and bosses, teachers and police, authorities of all types fill our civil society. Disobey one and suffer the consequences. But surely our values—the ethical rules we try to live by—would hold firm against abusive authority in a one-hour psychology study?

Think again. The single most important research project in the history of psychology was surely Stanley Milgram's famous series of experiments on obedience, run during the decade before My Lai. The chilling outcome: most of the ordinary people who volunteered delivered what they thought were painful and potentially dangerous shocks to another ordinary person, on the instructions of a researcher.[2]

The study was supposedly about the role of punishment in learning and memory. Picture two men showing up at a Yale laboratory: One, introduced as "Mr. Wallace," is actually a member of the research team, but the real volunteer doesn't know that. The volunteer and the accomplice Wallace draw lots to be "Teacher" or "Learner"; the drawing is rigged so the volunteer is always "Teacher." Mr. Wallace, the learner, is hooked up to a shock machine, where "Teacher" is given a mild shock so he'll believe that he will be giving Wallace real shocks. (He won't be; this was the only real shock in the study.) While "Teacher" is still in the shock room, Wallace points out that he has been diagnosed with a "slight heart condition." Both men are assured by the researcher that the shocks cause no permanent damage.

In the adjoining control room, "Teacher" reads word pairs to Mr. Wallace over a microphone and then tests his recall. Each time Mr. Wallace makes a mistake, "Teacher" follows instructions and presses buttons to give him shocks of steadily greater intensity. The scale goes from "Slight" to "Intense Shock" all the way to "XXX": 15 volts to 450 volts, in 15-volt increments.

Would the "teachers" follow instructions and give shocks up to the highest level? Alarmingly, in a pilot study where the learner made no audible complaint, almost none of the "teacher" participants protested to the researcher—in sharp contrast to predictions made by experts as well as ordinary people. In the "slight heart condition" experiment, the most-cited of the variations that Milgram ran, taped protests were played at set positions on the shock scale.

They included repeated cries to "Let me out of here!" and agonized screams. Nonetheless, two-thirds of the forty participants obeyed fully, going all the way to "XXX."[3]

Not without protest. But, if the "teachers" asked about injury to the learner, they were again assured that "Although the shocks may be painful, there is no permanent tissue damage, so please go on." Many participants simply abdicated responsibility to the experts in charge. The study was soon repeated around the world, and similar results have been consistently found.[4]

While it's easy to say that we would resist, in the actual situation, we have to accept that most of us would not. In a way, it's a type of self-control challenge: At a distance, the right choice is easy to make. But there in the hot seat, abstract values pale in comparison to the immediate negative consequences of defying authority—something most of us have been taught for years *not* to do. On a different scale, it's like setting your alarm clock for an early start with good intentions, then giving in to your overpowering sleepiness the next morning. Adding to the challenge, the obedience-research participants had no warning of what was to come. As one participant said, he felt "totally helpless and caught up in a set of circumstances where I just couldn't deviate and I couldn't try to help."[5]

Disobedience was a lot harder than it sounds, then. How much more so would it have been to defy an order to shoot in wartime and risk military justice, as at My Lai? And as Milgram pointed out, those who did the actual killing at the Nazi death camps were low in the hierarchy, obeying orders during a war with their own lives at stake.

People face diabolical choices in peacetime, too. At one of Australia's nineteenth-century penal stations, an informer system meant that convicts-turned-guards were flogged if they didn't report insubordination or minor problems. If they did report them, of course they were hated by the convicts. If they defied the system entirely, they were thrown back into the regular convict pool, suffering vicious reprisals.[6] Sometimes there are no good choices.

In the Asian practice of suttee, some women have been asked—or forced—to die on their dead husbands' funeral pyres. How must their children have felt, given the social, economic, and religious consequences of disobedience? Mahatma Gandhi not only condemned suttee, he also set the most famous example of civil disobedience on a vast scale—a movement that won India's independence.

The lesser obedience challenges most of us face are more likely to be sleazy doings at work. Should you look the other way? Or do the ethical thing and risk losing your friends, your job, and maybe your career? Societies that are serious about ethics support whistleblowers, giving them a fair hearing and protecting them from consequences like reprisal. And these societies teach their kids when *not* to obey.

OVERCOMING PREJUDICE

These societies also teach their kids to avoid prejudice. Sometimes that means disobeying community rules and refusing to conform, like Atticus Finch in the racist society of *To Kill a Mockingbird*. Not easy. Serious consequences.

In fact, prejudice facilitated the worst atrocities of the twentieth century, such as the Nanking Massacre, the Nazi death camps, and the Rwandan genocide. It's been with us probably as long as we could say "us" and "them," and it's particularly hard to change when it has the additional consequence of serving self-interest. For example, because slavery had become the basis of the nineteenth-century US Southern agricultural economy, the consequences of potentially losing it were dire for the ruling class. Racial prejudice helped to justify it.

The ins and outs of "us" and "them." Sex, age, disability, nationality —no one is immune, because a seemingly endless list of characteristics can be the basis for prejudice and the stereotypes that go with it. Even fashion: high schoolers who defy current styles can testify to this, if only to raised eyebrows and snubs. Conformity gets enforced by consequences, sometimes with good reason, sometimes not.

Psychologists have created real biases simply by giving randomly selected people different badges and assigning them to different groups.[7] It's part of the unwritten rules of our culture, "team spirit," if you will. Overgeneralizing from a few experiences, we can create our own stereotypes. More often, we learn them through modeling, social rules (spoken or unspoken), and the media.

Unexpectedly, even when we know we shouldn't, researchers find we have a tendency to believe what we read.[8] (It can be a mindless, automatic process flying below our conscious radar.) Or, as we saw in chapter 8, unobtrusively pairing neutral Pokémon figures with negative-emotion words and images instilled new, presumably classically conditioned emotional responses. Again, we don't always realize what's happening.

Fortunately, overwhelming the negative associations with positive ones is one way to fight back. There are plenty of others, too.[9] (For years, my car has sported the bumper sticker "Fight Stereotypes: We're All Individuals.")

Fighting back. Turnabout is fair play: a direct way to discourage prejudice is to experience it. In 1968 after Martin Luther King Jr.'s assassination, schoolteacher Jane Elliott got concerned about her third-graders' difficulty in imagining racial discrimination in nearly all-white Iowa. How could she teach them to walk in someone else's moccasins and take a different perspective? So she divided them based on eye color. One day blue-eyed children were given privileges like extra recess time and praise for their good behavior and intelligence, while brown-eyes were derided and restricted. The children joined in. The next day brown-eyes had it good and blue-eyes took their turn at the bottom—though the experience was less intense because these kids had been on the other side and the lesson was getting through.

Elliott repeated the exercise for each new class, and years later, her kids reported the impact of actually feeling the consequences of this very brief stint of discrimination. As one said, "Nobody likes to be looked down upon. Nobody likes to be hated, teased or discriminated against. And it just boggles up inside of you—you just get so mad." Another student said, "I felt demoralized, humiliated."[10] They thought the experience was valuable, and several studies suggest it does help reduce prejudice.[11]

Here's another direct approach: in a classic study from the 1950s, Muzafer and Carolyn Sherif randomly assigned white, middle-class, fifth-grade boys to two groups at a summer camp. The "Rattlers" and the "Eagles" competed against each other while the researchers made conditions rough, stoking the mutual dislike that developed. How then to achieve peaceful coexistence? The best method was to have the boys work together toward common goals. To rent a movie, for example, the Rattlers and Eagles had to pool their resources and agree on which one. It must have been fun to set up the next challenge: According to plan, a truck broke down on its way to get food for the hungry campers. They worked together to start it by pulling it with a tug-of-war rope. These strategies worked and the boys became friendly.[12]

As the Sherifs noted, consequences were critical. In the early stages with rough conditions, fighting over limited resources was a zero-sum game where some could win only when others lost. Not surprisingly, conflict arose. Later, when cooperation was the only way to get the consequences everyone wanted, cooperation became likely—along with some recognition of the redeeming qualities of "them."

Interacting with the immediate situation's signals and consequences were the boys' own histories and the cultural and individual rules they had learned. As we saw in chapter 11, some cultures reward cooperation more than others. In another classic study, Israeli children from a kibbutz and from a city played together, the kibbutz children having been taught a lifestyle of cooperation. Under consequences for groups as a whole, all the children cooperated. But when individual rewards became available, the urban children began competing (even though this strategy didn't pay off). The kibbutz children continued cooperating, true to their cultural training.[13]

Following up on these findings, social psychologist Elliot Aronson and his colleagues created the "jigsaw classroom" in which diverse youngsters work together toward a common goal (like puzzle pieces fitting together).[14] Reviews of this sort of research have found real benefits,[15] and many US elementary schools use some form of cooperative learning.

Getting to caring. Beyond an income that provides for reasonable quality of life, researchers find that more money often fails to create more happiness. Instead, happiness comes from rewarding social relations and self-fulfillment.[16] "Be happy to be good" is another consistent finding.

Perspective taking, an important part of fighting prejudice, also fuels altruism. Still, caring for others in distant countries or future decades is harder than caring for people we know—or at least can see. The famous "Baby Jessica" received hundreds of thousands of dollars in donations after she fell down a well in a blaze of publicity. Social-psychology researchers find that people contribute substantially more to save one child whose photo they can view than to save many equally deserving victims who are nameless statistics.[17] More emotion, more reward value—that's one reason (see chapter 8). Other reasons include learning that we can actually make a difference with just one person, whereas huge problems can seem too vast, like prejudice itself. (And a resulting too-lean schedule of reinforcement that doesn't offer sufficient rewards for helping can lead to "social worker burnout" in anyone, not just social workers.) Then there's the diffusion of responsibility that comes with being part of the bystanding masses (see chapter 11, "someone else will do something"). Getting to caring can be an uphill battle.

Still, we don't always require personal contact or photos. Indirect methods can foster both caring and prejudice fighting. How can we facilitate perspective taking, for example, discovering the hidden consequences motivating what strangers do and say? George Eliot said, "the only effect I ardently long to produce by my writings, is that those who read them should be better able to *imagine* and to *feel* the pains and the joys of those who differ from themselves."[18] Long before the prejudice-fighting blockbuster miniseries *Roots* (which aired in 1977), Harriet Beecher Stowe's bestselling antislavery novel, *Uncle Tom's Cabin*, was so successful that Abraham Lincoln, meeting her in 1862, reportedly said, "Is this the little woman who made the great war?"[19]

These prejudice-fighting efforts have been paying off. For example, starting from what now seems like an astonishing low of 4 percent in 1958 (the year I was born), US approval of interracial marriage has climbed to 86 percent. That's not just because new generations grow up with more modern ideas. Older people changed their minds as well.[20] There's hope for humanity yet.

POLITICS: THE ART OF THE POSSIBLE MEETS THE SCIENCE OF CONSEQUENCES

Watch the spectacle of politics throughout history, though, and no one could be blamed for wondering about that. *Cui bono?* Who benefits? And who pays the price? Even privileged absolute rulers like Alexander the Great and Elizabeth I had plenty of vested interests to cope with. Politics is about consequences in conflict.

These days, political and economic systems are intended to provide rewards that motivate hard work, innovation, and creativity (see chapter 1). But in a scene oft-repeated through history (and noted by Aristotle), short-term benefits for a politically powerful few can end up trumping all other considerations. Two examples from the nineteenth century follow.

British sailors who refused to serve on overloaded cargo ships could be (and were) sent to prison. However, unscrupulous ship owners made more money with heavier loads, and marine insurance meant they did not suffer financially when the ships foundered and sailors died—as many did. (An unanticipated consequence of insurance: in some cases, heavily insured "coffin ships" were worth more sunk.) It took years for reformer Samuel Plimsoll to navigate the politics and get a law passed prohibiting overloading.[21] A law with teeth that actually made a difference, that is. No accountability, no consequences, no change.

Similarly (to pick on the Brits again), when private British contractors transported convicts to Australia, the job was sometimes done minimally so there would be more profits for the owners. A large percentage of the convicts died under conditions similar to those in slave ships. In a successful political response, independent naval agents were placed on board (with accountability) and conditions turned around.[22]

The complete stories behind these reforms are as full of drama as any Shakespeare play, and political scholars put years of effort into understanding the changing, conflicting motivations of key leaders and vested interests. The hitch is that this complexity can rival that of the nature-nurture relations that provide the context for everything (see part 1). It's hard enough to understand the motivations of one person. Add a few others and suddenly you've got n-body physics. As far as consequences go, all the principles of parts 2 and 3 are acting—schedules, signals, delays, rules, and the rest—multiplied across people in a chaotic-seeming jumble across the years where a butterfly's wing-flap ends up making empires fall.

Indeed, when consequences conflict anything can happen, and sometimes obviously useful technologies get ditched. The Japanese received guns from Europeans in the 1500s and the initial response was enthusiastic. For a variety of reasons, however, the sword-fighting samurai in charge gradually outlawed them, leaving Japan vulnerable later.[23]

Some outcomes *are* predictable, though. Just as at the Sherifs' summer camp, limited resources frequently set off conflict: Zero-sum games are the hardest to play. An addition to all the other reasons for conflict in the Middle East, for example, is competition for increasingly scarce water.

Cooperation would seem the most rational solution for coping with limited resources, and more generally as well. Treaties like START (Strategic Arms Reduction Treaty) are cooperative attempts to prevent nuclear war, and that's in everyone's best interest. Mutual rewards smooth relations, national as well as individual: "You scratch my back, and I'll scratch yours" is an almost universal norm. But if there's a chance of winning . . . Not surprisingly, world politics results in networks of both cooperation and competition, interweaving and flip-flopping, trade agreements being a prime example.

The United Nations is another example of worldwide cooperation. It doesn't always succeed at its humanitarian goals, and experts still argue about how many UN troops might have been enough to stop the worst of the Rwandan genocide, which killed almost 1 million people. But there is little argument that preventing problems before they start is the ideal. It's just not always politically possible. In retrospect, World War I seems to some experts like an idiotic, readily avoidable war. But it happened.

Adding to the challenge, intended consequences don't always happen and completely unexpected consequences do. Sometimes the results are serendipitous. How did civilization develop in the first place? From unanticipated consequences. Prehistorically, in areas with wild plants and animals suitable for domestication, no one said, "I'm tired of hunting and gathering, let's switch to farming next year. And then how about inventing writing?" Instead, bands of people gradually shifted to farming because of the short-term rewards: more food security, for example. One thing led to another, shaped by consequences.

Then again, unexpected consequences can run the best-planned schemes off the rails. And how unfair it is when terrific short-term rewards surprise us with awful long-term negatives. The Sumerians in the Middle East's Fertile Crescent created the first civilization considered to have developed agriculture. They also discovered that irrigation eventually led to the buildup of problem salts in their soil, useless farmland, and the breakup of their society.[24] Their dismay has been repeated over the millennia to this day: witness my own home of California's Central Valley, where thousands of irrigated acres in marginal areas have been turned into wasteland of no use for crops or wildlife. We know better—we have science—but it can't help if we don't use it. Conflicting consequences, that's politics.

THE SHORT TERM VERSUS THE LONG TERM: HAVING IT ALL?

A perennial source of conflict stems from the lure of the immediate: the rewards of grabbing are always hard to resist. Delayed consequences pack a lot less punch, no matter how important they are (see chapter 12), and any uncertainty is an excuse to ignore them. Climate change, for example, is a global problem in which future benefits—like weather that doesn't seesaw out of whack—require unpleasant belt-tightening now. In other words, it's a case of short-term versus long-term *consequences*, requiring large-scale self-control. No wonder we are having trouble dealing with it despite the scientific consensus.

In his science fiction classic *The Gods Themselves*, Isaac Asimov imagined an Earth in which cheap, clean energy had become available, but at a rather significant delayed cost: the potential explosion of our part of the universe. Solutions that involved weaning everyone off the energy source were not politically feasible; as one of the characters sadly noted, "The easiest way to solve a problem is to deny it exists."[25] In Asimov's novel, a win-win solution fobbed the potential costs off to a different universe. Earth got to have it all.

But that's science fiction. Dealing with the long-term problems of the real planet will be a little harder.

Failures. There's no shortage of societal self-control failures to serve as warning shots across the bow. Environmental issues alone offer a wealth of examples:

- Overfishing, such as the loss of the rich Atlantic cod fishery (see chapter 12) and many other fisheries, ancient as well as modern.[26] I remember how shocked I was at unknowingly polishing off orange roughy that may have been older than my grandmother (who lived to be 100). Talk about unsustainable.
- Too much fertilizer. Hundreds of the resulting "dead zones" now afflict waters around the world, including thousands of square miles of the Gulf of Mexico. Because each individual farmer benefits from using ample fertilizer now, fisheries and ecosystems suffer later.
- Apartment builders and developers who skimp on insulation and energy-efficient appliances. They are expensive, and the increased running costs later don't affect their profits but the upfront costs do. (Homeowners are more likely to complain.)
- Overuse of antibiotics in people and farm animals. Antibiotic-resistant strains of bad bugs are flourishing as a result, and people are dying.
- Alien invasives like kudzu, fire ants, zebra mussels, and chestnut blight fungus. Around the world, accidental or intentional introductions

gone awry cost many billions of dollars each year, destroying econo-
mies, and devastating native ecosystems (see chapter 13).[27] An ounce
of prevention . . .

Many of these examples illustrate "moral hazard," a behavioral economics
term for what happens when people who take risks can potentially benefit while
avoiding suffering any negative consequences. It's been applied to the recent
Great Recession, for example. Over several decades, banks lobbied successfully
for less regulation. Risky speculation unbound helped create a lucrative but
ultimately disastrous lending bubble that burst in 2008. Ratings agencies
that were supposed to be part of the system of checks and balances failed.

Alas, history shows that it can take severe consequences like this to motivate change in business as usual. After the Great Depression, the financial
sector got sorely needed regulation. But at the same time, the US Food and
Drug Administration got nowhere attempting to improve regulation of drug
manufacturing—despite trying to publicize a number of resulting deaths. The
drugmakers lobbied against regulation as an unnecessary cost, and a media
dependent on drug-advertising revenue failed to help the FDA with publicity.
After the "Elixir" tragedy, in which an incorrectly prepared drug killed over
100 people, regulation came through almost overnight.[28] More recently, after
the British Petroleum Deepwater Horizon explosion and oil spill, the government commission concluded that insufficient regulation met poor safety procedures, and new regulations are in the works.

Sometimes the wake-up call comes too late for a whole society. Easter
Islanders failed to live sustainably, driving to extinction all their large trees,
all their land birds, and most of their seabirds. The consequences were cata-
strophic: a population crash and the loss of the ability to construct watercraft
large enough to sail to the nearest inhabited islands.

On our island earth, we need to make better choices.

Successes. Fortunately, we do have plenty of successes to give us hope.
When chlorofluorocarbons were destroying the atmospheric ozone layer that
barred harmful ultraviolet rays, a united effort instituted a ban, and better
alternatives were found. (I remember this effort distinctly, since I wrote a
college paper about it while it was going on.) In his book *Collapse,* Jared

Diamond pointed to the Tikopia islanders, who have achieved a sustainable culture for almost 3,000 years on just 1.5 square miles in the midst of the Pacific. Similarly, the Tokugawa shoguns in seventeenth-century Japan arranged a successful sustainability plan for their disappearing forests.[29]

The late Nobel Prize–winning economist Elinor Ostrom and her colleagues studied what's working these days. Considering all the competing consequences, it's not surprising that, as Ostrom noted, there are no universal answers. Instead, we are dealing with "complex, multivariable, nonlinear, cross-scale, and changing systems."[30] Solutions have to be tailored to fit.

For example, nationalizing forests sometimes helps promote sustainable use. But when a government can't enforce preservation, restricted access previously maintained by local people (acting in their own long-term interest) sometimes becomes open access for lawbreaking outsiders with no long-term consequences to restrain them. Ostrom cited Thailand, Nepal, India, and Africa as examples.

Community fishing cooperatives often see the rewards of sustainability and agree on restrictions and penalties ("let's keep this area off limits as a reserve"). Members who break the rules might be shunned. The evidence indicated that wasn't enough to be effective, though, Ostrom concluded. It took regular monitoring and the possibility of more serious consequences like fines. Cooperatives that have these capabilities can beat national governments.

But sometimes national governments do better. In Mexico, for example, the vaquita porpoise is the most endangered cetacean in the world, with fewer than 300 left in the wild. Under local management, it declined, an accidental "bycatch" victim from gillnet fishing. The government has stepped in and started paying boat owners to stop fishing or to use porpoise-friendly nets.[31] Will it be enough? Who knows, but at least it's a chance.

Experts like Cambridge economist Andrew Balmford have pored over the details, finding in Malaysia and Cameroon that sustainable logging produced significantly higher "total economic value." Maintaining mangrove habitat in Thailand instead of replacing it with shrimp farms had even higher benefits; likewise for *not* draining Canadian freshwater marshes in a productive farm zone. Sustainable fishing in coral reefs in the Philippines trounced dynamite fishing. The economic time horizons for these calculations varied from just 10 years to 100 for the Malaysian forests, and that's a big part of the problem.[32] What if you're starving today? Dynamite fishing doesn't look so bad.

In articles with titles like "Biodiversity Conservation and the Eradication of Poverty," experts try for some approximation of Asimov's win-win.[33] Sea turtles are all endangered now, but people in developing nations who used to rely on their eggs still need to live. Environmental-sustainability programs pay them not to poach turtle eggs, and people in the community often do the patrols and enforcement. Many of these sorts of programs are succeeding. But they're not cheap.

In an influential *Science* article, Balmford and his colleagues examined the financial value of preserving natural ecosystems. They concluded that the benefits-to-costs ratio came in at roughly 100 to 1: Spending $45 billion annually on preservation pays back about $4.4+ trillion annually.[34] (That's based on reams of data on actual logging, farming, fishing, carbon absorption, water filtration, soil preservation, flood prevention, and so on.) But are we willing to pay the price now? What about all our other problems?

There is no easy win-win, no science-fiction happy ending in sight (yet). But consequences will inevitably be part of the attempt.

Solutions: consequences to the rescue. Science has been our most reliable way of discovering rules that accurately describe the consequences to be expected down the line. Can we learn how to help these long-term consequences influence us now, even though immediate consequences are pulling us in a different direction? B. F. Skinner expressed the views of many when he called self-control humanity's "only hope."[35]

Education is where it all starts: it changes our everyday rules, helps us make individual long-term choices, and helps us reward our leaders for doing so on a grander scale. The media are an important means of spreading the word. But education and dissemination alone clearly aren't always enough.

Just as consequences help us achieve individual self-control (see chapter 12), they can assist with long-term planetary goals. A few examples of both small-scale and large-scale approaches to environmental sustainability include the following.

Adding/Subtracting Consequences

- Carpooling. High-occupancy lanes in freeways during rush hour reward making the effort to carpool to work.

- Supporting products like dolphin-safe tuna and the Forest Stewardship Council's sustainable lumber. As Jared Diamond put it, "Businesses have changed when the public came to expect and require different behavior, to reward businesses for behavior that the public wanted, and to make things difficult for businesses practicing behaviors that the public didn't want."[36]

- Energy use. In California, Washington, Minnesota, and several other states, people cut back on their energy use when their utility provided periodic reports featuring happy faces for households with less-than-average use, compared to similar neighbors.[37] It's easy to do and tends to result in a large benefits-to-costs ratio. (Don't want to boast, but I earned *two* smileys on my last statement.)

- "Cap and trade" markets for greenhouse gases, acid rain, and other pollutants. The idea is to create rewards for organizations that succeed in reducing their footprints.

- Water conservation. Homeowners fond of thirsty green lawns and exotic flowers in the desert have helped drain the rivers and aquifers of the American Southwest, to pick one example of a worldwide problem. Los Angeles paid residents $1 per square foot of lawn converted to drought-tolerant alternatives like native plants—better for the ecosystem, too, and likely to attract more wildlife.[38] Not to be outdone, Las Vegas offered $2 per square foot.[39] The resulting savings in water costs are another important reward.

- Biodiversity. Through the voluntary US Safe Harbor and Candidate Conservation Agreement programs, incentives make it a win-win situation for landowners to preserve endangered-species habitat.

Signals

- Energy use. From chapter 12, electricity meters can give immediate feedback, like the fuel-efficiency gauges in vehicles (remember the "hypermilers"?). Sometimes there are additional color cues, such as green for low use and red for high.

Schedules

- Littering. While they are always inherent, sometimes schedules are the focus themselves. In a high-litter area, adding variable rewards for properly disposing of trash succeeded so well that the schedule had to be made leaner.[40] (The reward was a free beverage.) Get people in the habit, then transition to the natural rewards of a healthier, litter-free landscape.

Commitment

- Fuel-efficiency standards. When we enact laws, we are committing to change. Similarly, we legislated lead out of gasoline, a worldwide success story with benefits both to the environment and to children's intellectual development.[41] (And crime as well, it's been suggested.[42] Schedules of reinforcement turned out to be a critical foundation of the legislation.[43])

Checklists and Charts

- Governments and businesses track progress. Just as for individuals, progress sustains motivation; lack of progress signals the need for more powerful consequences, new commitments, or other strategies for change.

Experience

- Recycling. People thought that Americans would never learn to recycle. It was viewed as too inconvenient. Wrong.
- Reducing. Manufacturers have long wanted to reduce their packaging—better for their bottom lines and for the environment, too. But more packaging can make it look like there's more product. Finally, people are learning that a smaller cereal box doesn't mean there's less inside.

Social Support and Rules

- Social norms combine these ubiquitous elements, in effect establishing informal rules enforced by conformity. (If everyone else puts out recycling, why don't you?) Tucson's new social norms for water conservation have made landscaping with drought-tolerant plants "cool." Water-conservation ordinances are more formal rules (but they don't always get enforced).
- Continuing with water conservation, social psychologist Robert Cialdini studied hotels offering the "green" option of not having towels and sheets laundered every day. To be most effective, the hotel's message had to indicate that most other guests reused their towels.[44] Social norms change consequence value even when no one's in the room.

Models

- Movie star Robert Redford and author Bill McKibben are well-known international environmental models.

How about dealing with all the problems we're facing right now? As nature-nurture systems experts Eva Jablonka and Marion Lamb pointed out, "If we want to solve 95 percent of the health problems in the world, what we need to do is give people enough to eat, and make sure they can drink clean water and breathe clean air."[45] We're still a long way from that, let alone the more ambitious set of eight short-term UN Millennium Development Goals, which all the UN member states agreed upon.[46]

Nobel Prize–winning economist James Heckman found in general that programs aimed at reaching at-risk children and preventing problems from developing in the first place often pay off handsomely for society at large, simply from an economic standpoint.[47] Science—including the science of consequences—has shown us how immensely flexible and resilient people can be, given a chance. Maybe someday, sooner than we think, we'll be able to give everyone enriched environments, so they can achieve their full potential—and contribute to more successful solutions.

Consequences are everywhere. . . . They are unavoidable, for good or ill. Can we learn to use them wisely and compassionately?

Each generation starts with hope that life will improve around the world. As it ages, each generation must accept that we still have a long way to go. But we can learn. We can learn to disobey when necessary. We can learn to care. We can learn what works and what doesn't, and change what we do. And we have.

When my dad was born in 1917, women couldn't even vote in most nations, including the United States. Lynching was a regular activity in the American South. The passenger pigeon, once the world's most abundant bird, had just gone extinct (and not a lot of people cared). And World War I was raging, with another even more horrendous to follow. Now, a century later, while there is no shortage of problems, we have made immense progress: less poverty and starvation in many nations (proportionally, anyway), more literacy, less prejudice, more environmental awareness. We may even share a general understanding worldwide of the need to try for future sustainability, with better quality of life for everyone.

Consequences will be part of humanity's grandest endeavor. How can we best use what we know? Skinner said: "Regard no practice as immutable. Change and be ready to change again. Accept no eternal verity. Experiment." And when we fail, "The real mistake is to stop trying." Let's keep learning. Let's never stop trying.

GLOSSARY

This is a selective, nontechnical glossary of the basic terms used in this book.

automatic behavior: A behavior performed with no or minimal conscious awareness. Example: An experienced driver can brake at a red light while deep in conversation. See chapter 9.

aversive: See **negative**

chain, behavioral: A series of behaviors performed in order, often (although not necessarily) automatically. Example: toothbrushing.

classical conditioning: A form of learning built around reflexes and similar unlearned reactions. In the salivation reflex, an "eliciting stimulus" like meat produces salivation. Pairing an eliciting stimulus with a neutral stimulus (such as an electric bell) produces classical conditioning when the neutral stimulus acquires the ability to elicit the reflex, becoming a "conditioned eliciting signal." See chapter 8.

clicker training: One of the positive-reinforcement-based forms of animal training. The clicker is paired with other rewards so that its sound becomes a learned reward. It can then be used as an immediate reward to teach a behavior or a signal. Once behaviors and signals are learned, the clicker isn't needed any more. See chapter 13.

consequence: An outcome that depends on a behavior. Example: the behavior of turning your head provides the consequence of looking out the window and seeing a view.

consequence, natural: A consequence that occurs normally in "real life." Example: Natural rewards for exploration include a variety of novel views. Similarly, learning how to read offers natural rewards.

delay discounting: The value of delayed consequences is frequently reduced or "discounted," compared to the value of comparable immediate consequences. See chapter 12.

epigenetics: A term that means "associated with genetics." Genes code for proteins, but the proteins are produced only when the genes are activated and transcribed. Within the cells, epigenetic mechanisms influence gene-activation patterns; that is, they help determine whether a gene produces its protein or not. See chapter 3 for a lot more information on this technical subject.

extinction (classical conditioning): After classical conditioning, a neutral stimulus (such as the electric bell for Pavlov's famous dogs) can become a conditioned eliciting signal. In extinction, repeated presentation of the conditioned eliciting signal without the original unlearned eliciting stimulus means the reflex stops occurring. Example: The bell is repeatedly heard without any meat available. As a result, salivation stops occurring in response to the bell.

extinction (schedule): The discontinuation of the consequences for a behavior. Example: You want your dog, Spot, to stop begging, so you simply stop ever rewarding begging.

functional analysis: The systematic determination of the factors actually influencing a behavior. Example: a youngster who kept coughing for months. Medical explanations were ruled out, and researchers investigating the possible role of consequences eventually showed that parental attention was rewarding and thus maintaining the coughing. (For more on this example, see chapter 9.)

generalization (behavioral): The transfer of function to similar behaviors. Example: Elementary school students learn handwriting from models and are rewarded for copying them. However, they also learn they can vary the way they shape their letters to some degree and still produce legible handwriting (accepted by teachers and others).

generalization (signal): The transfer of function to similar signals. Example: a toddler who correctly identifies capital *A*'s in different font styles than the one she learned from. In an "overgeneralization" example, a toddler starts calling the family cat as well as the family puppy a "doggie."

matching law: A mathematical equation that describes individual choice patterns when simultaneous behavior options are available, each with a different schedule of consequences. The matching law works across a wide range of species (including people), consequences, and behaviors. In its extended form, the equation includes the effects of delay, behavioral effort, different consequence amounts, the clarity of associated signals, and so on. See chapter 7.

meta-analysis: A systematic statistical approach to combining results from a large number of different experiments.

negative: If a behavior declines because of a consequence, that consequence is a negative (a punisher, an aversive) and the relationship is punishment. (Note that the negative can be an outcome involving the removal of a positive; see chapter 6. Other causes for the behavior, such as Pavlovian eliciting stimuli, must be ruled out.) Example: Swim in Lake Superior in June. Brrr! Never do it again. It's been punished by a negative, the cold temperature.

It's important to note that things that seem like negatives sometimes aren't: what matters is what actually happens, not the intention or appearance. This book gives examples of events that seem like they ought to be negatives but don't weaken the behavior they depend on (see, for example, chapter 1). Such "negatives" that have no effect on a behavior are *not* truly negatives; indeed, if the behavior is strengthened, they are reinforcers. Again, it's what actually happens that matters.

punisher: see **negative**.

punishment: see **negative**. Put simply, the opposite of reinforcement.

reinforcement: When a consequence sustains a behavior, the relationship is reinforcement. (Other causes for the behavior, such as Pavlovian eliciting stimuli, must be ruled out.) The consequence can be an outcome involving the presentation of a positive or the removal of a negative, or both simultaneously (see chapter 6). Simple positive example: You look out the window of your new hotel room and see an attractive view. As a result, you look out periodically throughout your stay. Looking out has been reinforced. Simple negative example: Whenever you see dirt on your rug, you clean it up to escape the unsightliness.

Technically, the full definition of reinforcement is more complicated, as relative probabilities, "behavior classes," and multiple aspects of behavior in addition to its rate (such as duration or intensity) can be involved. (This is also the case for punishment.) A good discussion is in Catania, 2006, chapters 5 and 7.

reinforcer/reward: In this book, these two terms are used interchangeably. Reinforcers depend on behaviors and sustain them as a result of this relationship. (Other causes for the behavior, such as Pavlovian eliciting stimuli, must be ruled out.) If a behavior gets going and/or keeps going because of a consequence, that consequence is a reinforcer.

It's important to note that things that seem like reinforcers/rewards sometimes aren't: what matters is what actually happens, not the intention or appearance. This book gives examples of events that seem like they ought to be reinforcers but don't strengthen or maintain the behavior they depend on (see, for example, chapter 1). Such "rewards" that have no effect on a behavior are *not* truly rewards; indeed, if the behavior is weakened, they are negatives. Again, it's what actually happens that matters.

reinforcer, generalized: A reinforcer that is broadly effective because it is associated with a number of other reinforcers. Example: money.

schedule of consequences: In short, the nature of the dependency of the consequence upon the behavior. Looking out the window always brings a view, but most of the time, the relationship between behavior and consequence is more complex. Paychecks require a lot of work, for example. Schedules can be work-based, time-based, duration-based, fixed, variable, or some combination, simultaneously or alternately available, signal-dependent, and many other permutations. Example: On a variable work-based schedule like reading the newspaper, perusing the pages brings periodic but unpredictable rewards. See chapter 5.

selectionism: A form of causal relationship in which variations occur over time, and success is later reproduced, while failure isn't. It's how learning from consequences works (in parallel with natural selection in evolution). Example: As you learn how to make free throws in basketball, your weak tosses don't get reproduced while your stronger, more successful tosses do. The stronger tosses have been selected via reinforcement. See chapter 2.

shaping: The process of reinforcement of successive approximations to a target behavior. Example: Kids shaping a quiet teacher into speaking loudly pay attention and smile whenever his voice is raised—at first only slightly, then, gradually building on the changes, only when he's close to a shout. See chapter 11.

signal (classically conditioned): A "conditioned eliciting stimulus" for a reflex. Example: An electric bell associated with meat becomes a classically conditioned signal that itself elicits salivation—it is now a conditioned eliciting signal. Before the bell was associated with meat, it was a "neutral stimulus" that did not elicit salivation.

signal (consequence-based): A feature or event that is associated with an increased (or decreased) chance of consequences. Example: Kids learn that swearing gets rewarded among their pals but punished near parents or teachers. In this book, I also occasionally use "signal" to mean a cue that modifies consequence value. Example: to enhance their motivation, dieters sometimes post on their refrigerators pictures of themselves when they were thinner.

systems approach: Taking the whole shebang into account, as necessary. That means all factors that can affect living things, from epigenetics to reinforcement histories to social norms.

timeout: A pause in the current schedule of positive reinforcement. A mild negative, it has become the preferred alternative to spanking for many parents and parenting experts. Example: Children who misbehave while playing have their favorite toys briefly removed. See chapter 11.

ACKNOWLEDGMENTS

From the beginning, this book benefited from the support of two of my colleagues in particular: Paul Chance and Richard Shull. My gratitude to them runs long and deep. Both of them also provided thoughtful comments on many of the chapters. Another stalwart supporter has been Karen Pryor, whose popular books helped inspire this one.

I began the research for this book at West Virginia University, and I am grateful to the Psychology Department for hosting me as a visiting scholar. Most of the actual writing got done while I was a visiting scholar at the University of the Pacific, and likewise, I'm grateful to the Psychology Department there. In particular, my colleagues Matt Normand and Carolynn Kohn have been wonderful sources of support and intellectual stimulation, and both of them provided helpful comments on chapters.

I am grateful to the generous friends, colleagues, and family members who took the time to review chapters: John Baldwin, Sara Blauman, Jewell Brown, Clarissa Bush, Sergio Cirino, Flavia Filimon, Pam Fish, Glen Fjelstrom, Sigrid Glenn, John Hall, Pat Hammer, Chris Harshaw, Sharon Jarvis, Kent Johnson, Bob Lickliter, Cathy Mathis, Mark Mattaini, Tim Miller, Ed Morris, Roy Moxley, Allen Neuringer, Dave Palmer, Alliston Reid, Dave Schaal, Kathy Schick, Hank Schlinger, Charles Spurr, Cathy Williams, and Joanne Schneider Willman. And I'm grateful to those who helped in other ways: Tom Critchfield, David Moore, Chris Newland, Stuart Vyse, and Criss Wilhite.

If I've forgotten anyone who helped, I apologize. I also take responsibility for any mistakes or errors of omission in the book. And naturally, those who helped don't necessarily agree with everything in it.

A big thank you to my agent, Laurie Abkemeier, who's been terrific to work with. She's improved the book in many ways. Thanks to my editor, Steven

L. Mitchell, for his guidance and support throughout the process. Thanks to artist extraordinaire and fellow naturalist Rene C. Reyes for agreeing to do the illustrations.

The support of my family has provided my daily consolation while I've been chained to my computer. My mother and my brother Ken devotedly read every chapter and supported me and the book in too many ways to name. My dad did not live to see the completion of this book, but I will always remember his excited "Go for it!" when I consulted him about whether to attempt it. I know he would have been pleased that the long journey has achieved its goal.

Finally, I'm grateful to all the scientists who've worked hard to build the research base. I'm grateful to have been a small part of that grand endeavor myself. And I'm grateful to have known pioneering psychologist B. F. Skinner as a friend as well as a professional colleague. One of the links that brought us together was our shared concern for the problems of this planet and its people. I hope this book can help us work toward a bright, sustainable future.

NOTES

This book provided an opportunity to attempt to pull many fields together as they related to the science of consequences: animal and human learning, language, and cognition; evolutionary biology; neuroscience; developmental psychobiology with its insights on nature-nurture relations; and the larger realms of biology, psychology, behavioral economics, education, and many allied sciences. To build what I hoped was sufficient knowledge of the immense research base, I spent over a decade reading literally thousands of scientific books and journal articles. It was a fascinating trip, and I hope that comes across in this book. Space limitations meant I've had to pick and choose what to cover very carefully, and, inevitably, I wish I could have included far more. This book represents my own views after careful weighing of the scientific evidence. I have tried to be balanced and inclusive, as far as the limited space allowed.

In addition to the references in this book, more references, extensions, and applications can be found on the book's website, www.scienceofconsequences .com. Please let me know of mistakes via the website so I can correct them. Because reference citations are provided in full in the bibliography, these notes just give sufficient author-date information for readers to find the correct citation.

PREFACE

1. It's really a misnomer to say "*the* science of consequences" because many fields of study rely on consequences. However, the science founded by B. F. Skinner, a primary focus of this book, arguably has the best claim: It was the first to establish an interlinked set of scientific principles applying to all consequences. It's been a launchpad for the other areas. And it's become a mature, mathematically sophisticated science

in its own right, now with extensive neuroscience supporting the behavioral relationships. What's more, it's created reams of evidence-based applications. I've tried to at least touch on most of the associated interdisciplinary areas, and to integrate all discoveries and applications in an inclusive, balanced way.

Behavior analysis has become the name for the academic field specializing in the basic and applied sciences of consequences per se as its core specialty area. The term for the process itself is *operant learning*, with older alternative versions being *instrumental learning* or *operant conditioning*. Pavlovian processes (discussed in chapter 8) have alternative names too: *Classical conditioning* is probably the most common, but also regularly used are *respondent conditioning* and *Pavlovian conditioning*.

PART 1: CONSEQUENCES AND HOW NATURE-NURTURE *REALLY* WORKS

CHAPTER 1. CONSEQUENCES EVERYWHERE

1. Passarelli et al., 1999; Wells, 1967.

2. This definition thus eliminates any eliciting effect that the putative consequence might have, such as in a Pavlovian process (see chapter 8). The full technical definition of *reinforcement* and *punishment* is actually considerably more complex, dealing (for example) with relative probabilities. See Catania, 2006, chapters 5 and 7.

3. See examples in chapter 14.

4. Durrell, 1956, p. 53. More recently, a YouTube video circulated of a sulphur-crested cockatoo "dancing" to the beat of a Backstreet Boys tune. Researchers studying this bird were convinced it was following the rhythm: Patel et al., 2009.

5. Porter and Neuringer, 1984.

6. Java sparrow: Watanabe and Nemoto, 1998. Rats: Cross, Halcomb, and Matter, 1967.

7. Eiders: Roberts, 1934. Starlings: Harwood and Porter, 1977. Swifts: Teale, 1948. There's no shortage of these sorts of examples.

8. Brennan, Ames, and Moore, 1966.

9. Barnes and Baron, 1961.

10. Meyer, 1968.

11. Griffin, 1931, p. 112.

12. Butler, 1954.

13. Richmond and McCroskey, 1995, p. 105.

14. Provost, 1990, p. 56.

15. Peterson, 2001, p. 37.

16. Pryor, Haag, and O'Reilly, 1969.

17. Goetz and Baer, 1973.

18. Page and Neuringer, 1985. Also see Neuringer, 2004.

19. Neuringer, 1986.

20. Bafile, "Reward Systems That Work: What to Give and When to Give It!" *EducationWorld*, updated June 25, 2012 (original work published 2003); http://www.educationworld.com/a_curr/curr301.shtml (accessed August 6, 2012).

21. Pryor, 1999, p. 3.

22. Carlstead, 1996, p. 328.

23. Harlow, Harlow, and Meyer, 1950.

24. Pryor, 1973.

25. Jolly, 1985, p. 148.

26. Least Heat Moon, 1982, p. 426.

27. Skinner, 1979, p. 282.

28. Widely attributed to Pauling.

29. Kish and Barnes, 1961.

30. Skinner, 1979, p. 293. His daughter was about nine months old at the time.

31. Similarly, manatees have been observed poking at blue crabs, apparently just to get a reaction (Sleeper and Foott, 2000, p. 78).

32. Carlstead, 1996, p. 322; Kavanau, 1963.

33. Catania and Sagvolden, 1980; Cerutti and Catania, 1997.

34. British job-control benefits: Marmot et al., 1978; Marmot et al., 1991. Also see, for example, Dunlap et al., 1994; Kern et al., 1998.

35. Langer and Rodin, 1976; Rodin and Langer, 1977.

36. Carlstead, Seidensticker, and Baldwin, 1991.

37. Pryor, 1999, p. 175.

38. The fact that Goldie sometimes initiated the game shows that this was not merely an elicited effect of my long eyeblinks, like meat eliciting salivation in Pavlov's dogs.

39. Sethi-Iyengar, Huberman, and Jiang, 2004. Also see Iyengar and Lepper, 2000.

CHAPTER 2. CONSEQUENCES AND EVOLUTION

1. *Instincts* are more technically known as *species-typical behaviors*. Filial imprinting and courtship rituals are just two of many examples. Instincts normally develop without the involvement of learning from consequences or Pavlovian conditioning, and hence can be described as "unlearned." Nonetheless, their development still entails the full slate of interacting nature-and-nurture factors.

2. Kessel, 1955. Courtship rituals in birds like manakins are also helpful in tracing evolutionary relationships.

3. Gottlieb, 1997.

4. Peterson, 1960; Bateson and Reese, 1969.

5. Harshaw, Tourgeman, and Lickliter, 2008. Individual chicks that heard the Japanese call independent of what they did were "yoked" to individual chicks that heard the Japanese call only as a consequence. In that way, both chicks heard the Japanese call at exactly the same times during the session, either as a consequence or not as a consequence. This is a standard way of separating the effect of simply presenting an event from its effect as a consequence.

6. ten Cate, 1994; also see, for example, Anderson, 2009.

7. King, West, and Goldstein, 2005.

8. Tumer and Brainard, 2007. More recent research has shown that these Bengalese finches can modify their song both upward in frequency in one location and downward in another, in response to a more complex reinforcement contingency. What's more, because of city noise, urban birds are modifying their songs: Dowling, Luther, and Marra, 2012.

9. Collias and Collias, 1964.

10. Avital and Jablonka, 2000, pp. 63–68.

11. Weiss, 1997.

12. Weiss, 2004.

13. Pessotti, 1972.

14. Brembs and Heisenberg, 2000.

15. Lee, Clancy, and Fleming, 1999.

16. Wilsoncroft, 1969.

17. Waddington's research is described in Avital and Jablonka, 2000, pp. 317–25. Waddington also described the same sort of genetic assimilation with a four-winged version of the fruit fly rather than the normal two-winged version.

18. Grant and Grant, 2003. Sympatric speciation relies on "assortative" mating—the greater likelihood that small-billed finches will breed with each other rather than with other finches. The woodpecker finch, which frequently uses tools, is a particularly striking example of behavior-led speciation. But adaptive radiation doesn't always happen even given substantial foraging differences (Werner and Sherry, 1987). As always, all the elements of the complex nature-nurture system interact.

19. Morin et al., 2010.

20. See, for example, Riesch et al., 2012.

21. Jablonka and Lamb, 2005, pp. 294–96.

22. Skinner, 1981b.

23. Grant, 1999.

24. Raine and Chittka, 2008.

25. Ginsburg and Jablonka, 2010. This reference is also a good source in general on the evolution of learning from consequences.

CHAPTER 3. GENES AND CONSEQUENCES

1. DeLange et al., 1969.

2. Eric Lander in interview "Meet the Decoders" accompanying transcript of PBS *NOVA* video *Cracking the Code of Life*, http://www.pbs.org/wgbh/nova/genome/deco_lander.html, updated April 2001 (accessed January 21, 2012).

3. Carroll, 2005a, p. 269.

4. Estimates go as low as 18,000 genes, as in Zimmer, 2008, p. 117. Alternative splicing is now known to be more common than originally thought, so one gene can be read to produce different (albeit related) proteins.

5. Bateson, 1988; Johnson, 2007, p. 169.

6. Carroll, 2005a.

7. Morange, 2001, pp. 88–89.

8. Gottlieb, 1998.

9. Mataga et al., 2001.

10. Zimmer, 2008, p. 190.

11. Pilegaard, Saltin, and Neufer, 2003.

12. Glaser et al., 1993.

13. Anokhin and Rose, 1991.

14. Kleim et al., 1996.

15. Yin et al., 1994.

16. Rapanelli et al., 2010.

17. Red or white eye color in *Drosophila melanogaster*, the model fruit fly, has been known for many years to be a sex-linked genetic characteristic—but the general "it's a system" principle holds.

18. Swann, 1999.

19. *Incomplete penetrance* technically describes the degree to which having the problematic genotype doesn't guarantee getting the disease. Genes are interacting among themselves as well as with all the other variables in the system. For an excellent discussion with examples, see Morange, 2001, chapter 4, pp. 48–63. Sickle-cell anemia is discussed on pp. 50–51.

20. Jablonka and Lamb, 2005, pp. 61–62.

21. Ibid., p. 63.

22. Cierpial and McCarty, 1987.

23. Suomi, 2002.

24. *Heritability* does *not* mean the degree to which a trait is inherited. It's defined as the proportion of the variation in a trait that is correlated with genetic variation. It applies only to a specific population under specific circumstances, never to individuals—and for the same trait in the same population under different circumstances, it can change dramatically, as the examples show. David S. Moore offers an excellent discussion in *The Dependent Gene: The Fallacy of Nature vs. Nurture*, 2001.

25. Lewontin, 1970; Moore, 2001, chapter 2.

26. Moore, 2001, chapter 3.

27. Shenk, 2010, p. 67.

28. Gilbert and Jorgensen, 1998.

29. Mice: Tamashiro et al., 2002. Cats: Zimmer, 2008, p. 49. Epigenetic contributions are considered likely.

30. Schulz et al., 2006.

31. Beisson and Sonneborn, 1965.

32. Fraga et al., 2005. Relatedly, *genomic imprinting* is an epigenetic mechanism by which the gene form (*allele*) of either the mother or the father is preferably expressed in the offspring—something that for years had been considered impossible. Humans have a number of imprinted genes.

33. Lyko et al., 2010.

34. Waterland and Jirtle, 2003; Morgan et al., 1999.

35. Heijmans et al., 2008.

36. Jablonka and Raz, 2009.

37. Licking the pups' salty urine appears to be reinforcing for the mother: Gubernick and Alberts, 1983. Male pups get more licking because the extra testosterone in their urine makes it more rewarding: Moore, 1992.

38. Roth et al., 2009.

39. Weaver et al., 2004.

40. Francis et al., 2002.

CHAPTER 4. NEUROSCIENCE AND CONSEQUENCES

1. Chapin et al., 1999; Carmena et al., 2003.

2. Hebb, 1949, pp. 298–99. Among themselves, rats have a characteristic greeting in which they close their eyes, yawn, lean forward, and stretch out their front paws—it's really cute. I was honored when my pet rat, Clover, started giving me this greeting from afar when I entered a room.

3. Lazic, Schneider, and Lickliter, 2007. That wasn't my first enrichment

project: one of my earliest studies (Schneider, 1988) showed benefits for larger rat laboratory cages with solid bottoms and bedding material, as compared to small, barren wire cages. That research line has taken off, and a recent summary showed that mice find such "enriched" cages more rewarding and will work to gain access to them.

4. Pinaud, 2004; Markham and Greenough, 2004.

5. Cory-Slechta et al., 2009. Also see chapter 5 on reinforcement schedules in this book. A fixed-interval schedule was used in this study.

6. Tang, 2001.

7. Gould, 2007.

8. Ibid.

9. Kleim et al., 2002.

10. Markham and Greenough, 2004, p. 360.

11. Stein, Xue, and Belluzzi, 1993.

12. Platt and Glimcher, 1999. Also see Louie and Glimcher, 2010.

13. Platt quoted in Kast, 2001, p. 128.

14. Hollerman and Schultz, 1998.

15. Many of these neurotransmitters are also found in simple invertebrates, sometimes serving similar functions. For example, cAMP has long been known to be involved in learning in two standard lab-model invertebrates, the fruit fly *Drosophila* and the sea slug *Aplysia*.

16. Dodd et al., 2005, quotes on pp. 1378–79.

17. Amphetamines as rewards: Kennedy, Caruso, and Thompson, 2001.

18. Kusayama and Watanabe, 2000.

19. Wise, 2004.

20. Xi and Stein, 1999.

21. Addolorato et al., 2007.

22. Berridge and Robinson, 1998.

23. Kelley and Berridge, 2002, p. 3307. Indeed, from what we now know about neural networks, some experts conclude that multiple function is the common rule.

24. Eisenberger, Lieberman, and Williams, 2003.

25. Music: Blood and Zatorre, 2001. Beautiful faces: Aharon et al., 2001. Money: Breiter et al., 2001; Knutson et al., 2003. Also, for social rewards, see Behrens et al., 2008.

26. Olds and Milner, 1954.

27. Summary chapter, including citations for goldfish, humans, and iguanas: Shizgal, 1999. Dolphins: Lilly and Miller, 1962. Newborn animals: Bacon and Wong, 1972. Snails: Balaban and Chase, 1989.

28. Hoebel, 1988.

29. Steiner, Beer, and Shaffer, 1969.

30. Deutsch and Howarth, 1963, p. 447.

31. Berridge and Robinson, 1998.

32. Kelley and Berridge, 2002.

33. Wise, 2004; Kelley and Berridge, 2002; Silva et al., 2007.

34. For example, Johansen et al., 2009.

35. Merzenich, quoted in Doidge, 2007, p. 56.

36. Recanzone et al., 1992 ; Recanzone, Schreiner, and Merzenich, 1993.

37. Merabet et al., 2008.

38. Amedi et al., 2008.

39. Sharma et al., 2000. Such flexibility is becoming a general finding. For a striking example, see O'Kane, Kensinger, and Corkin, 2004. Also, note people with only one brain hemisphere who can live reasonably normal lives (see chapter 10).

40. Taub et al., 2006.

41. Johansen-Berg et al., 2002.

42. Wolf et al., 2008. In a randomized study, participants are assigned randomly to different groups in an experiment.

43. Wolf et al., 2010.

44. Chapin et al., 1999.

45. Carmena et al., 2003.

46. Hochberg et al., 2006. For additional applications, see, for example, Brumberg and Guenther, 2010.

PART 2: THERE'S A SCIENCE OF CONSEQUENCES?

CHAPTER 5. CONSEQUENCES ON SCHEDULE

1. Vyse, 1997, p. 4. Vyse's book provides an excellent and accessible summary of research on superstition, including an important superstition experiment by B. F. Skinner in 1948.

2. Wagner and Morris, 1987.

3. Ono, 1987.

4. Hoffman, 1996.

5. Iversen and Mogensen, 1988. For more, see the chapter 12 discussion of this study. Also, Skinner (1938, pp. 67–69) found that most of seventy-eight rats tested quickly repeated a lever press after it had first been followed by reinforcement.

6. Higgins, Morris, and Johnson, 1989. For more on how rules interact, see chapter 10.

7. Grescoe, 2008.

8. Ferster and Skinner, 1957. Note that a single unreinforced behavior on a lean schedule can appear to be occurring for no reason. Stepping back and checking for the history of reinforcement is essential.

9. Mazur, 2005, p. 147.

10. Ibid., pp. 360–62.

11. Ward, 1976, quote on pp. 149–50. Also see Van Houten and Nau, 1980.

12. Petry et al., 2005.

13. Alan B. Christopher's (1988) unpublished doctoral dissertation, as described in Chance, 2003, p. 376.

14. Mazur, 2005, chapter 7. The fixed-interval "scallop" is the gradually increasing rate of responding as the interval end approaches. It doesn't always happen: for example, people sometimes time the intervals and respond only when the interval is over. Given a "clock" signal, such as a gradually disappearing bar, animals show this pattern, too. Without a clock, humans have been shown to be more likely to produce a nice scallop if their timing is distracted or if they're more motivated. Other factors are response effort and self-generated rules (see chapter 10).

In interval schedules, just waiting brings closer the opportunity for reinforcement. That's not the case in work-based ratio schedules. Note that time-based interval schedules still require work.

15. For example, McDowell, Bass, and Kessel, 1993; Li, Krauth, and Huston, 2006. For other applications see, for example, Marr, 1992.

16. There are a number of different types of low-speed schedules.

17. The schedule is called "differential reinforcement of low rate." For pigeon delaying tactics on a related schedule, see Fetterman, Killeen, and Hall, 1998. For the two-and-a-half-year-old children, see Pouthas, 1981. Of the many other schedule types, another worth a mention is the duration schedule, based on how long a behavior lasts—piano or basketball practice, for example. And "adjusting schedules" change as a function of our own behavior.

18. Weiner, 1964. In the reverse direction, Wanchisen and colleagues found that rats don't show the standard fixed-interval scallop when they're put on that schedule after they've had variable ratio experience: Wanchisen, Tatham, and Mooney, 1989. Also see Tatham and Wanchisen, 1998.

19. Feltovich et al., 2006, p. 45.

20. Marley and Morse, 1966.

21. For example, Lowe, Beasty, and Bentall, 1983. More generally, there is a huge literature on schedules of reinforcement across species. Typical schedule patterns are seen in invertebrates like octopuses, through the range of vertebrates, to people. For two examples using unusual species, see Grossman, 1973 (honeybees) and Scobie and Gild, 1975 (goldfish).

22. Wodehouse, 1959, p. 26.

23. John Hintermister, quoted in Connor, 1988, p. 132.

24. Skinner, 1968, p. 159.

25. Dews, 1955. There's actually a large literature on this subject. A good summary of several classic articles is in Branch, 1991. An example from the recent literature is McMillan, Li, and Hardwick, 2001.

Other drug studies have found the same schedule effects regardless of whether the consequence for pecking was food or avoiding a negative; see chapter 6 notes.

Schedule performance offers a sensitive baseline from which to investigate effects of drugs, toxins, and many other factors. The fixed-interval schedule, for example, has helped scientists set safe limits of exposure to lead (chapter 16).

26. Similarly, after construction of the Glen Canyon Dam made access to a noted natural feature trivially easy, Edward Abbey said it well in *Desert Solitaire*: "Half the beauty of Rainbow Bridge lay in its remoteness, its relative difficulty of access" (1968, p. 217). What's reinforcing on one schedule of reinforcement may not be on a different one. In both gambling and fishing, winning every time is thus not necessarily better.

CHAPTER 6. THE DARK SIDE OF CONSEQUENCES

1. Least Heat Moon, 1982, p. 51.

2. Montaigne, 1877.

3. Shigemitsu, 2005.

4. I think most teachers can relate. The following article gives an example of students' reactions to bonus points for on-time attendance: Perone, 2003.

Escaping a negative is a reinforcer—and the process is rather confusingly called negative reinforcement. It's "reinforcement" because it still strengthens. It's "negative" because something (a negative) is withdrawn. In the same way, "negative punishment" means withdrawing a positive. I will avoid these terms.

5. Schlund, Magee, and Hudgins, 2011; Niznikiewicz and Delgado, 2011.

6. Baron, 1991. Also see Abramson, 1986; Cook and Catania, 1964; Schneider and Lickliter, 2010b. There are some exceptions, however: Roberts et al., 2008.

7. This comparison-based effect on value is known more broadly as *relative deprivation*.

8. Perone, 2003, p. 3.

9. Muir, 1954 (original work published in 1909), p. 294.

10. Decker, 1996.

11. An aversively loud noise can become a reward: Ayllon and Azrin, 1966.

Relatedly, it's been suggested that one of the reasons some battered spouses stay with their partners is because of occasional rich rewards for doing so—and hope that the negatives will lessen or not occur again. In effect, variable positive and negative schedules are competing with each other, as well as interacting with signals, rules, histories, and the rest of the systemic factors.

12. Holz and Azrin, 1962.

13. Either habituation or its opposite, potentiation, can occur upon the repeated presentation of certain stimuli, depending on factors like stimulus magnitude and timing. Potentiation results in a greater response, rather than a diminished response. Pavlovian conditioning is related to habituation and potentiation; see chapter 8.

14. On a smaller scale, that big-screen TV is immensely exciting at first, but it soon becomes business as usual. Diener, Lucas, and Scollon, 2006.

15. Ulrich, 1966.

16. Maier, Anderson, and Lieberman, 1972.

17. Cherek and Pickens, 1970. In ratio schedules, when the ratio size is signaled, the size of the upcoming ratio is a major influence on the level of aversiveness—and on the likelihood of taking a longer break. Not surprising. See Perone and Courtney, 1992.

18. Goldman, Coover, and Levine, 1973.

19. Azrin, 1961. By pecking a timeout key, pigeons chose to spend up to 50 percent of their time off the fixed-ratio 200 schedule. Animals are also more likely to be aggressive during this period: Gentry, 1968.

20. Azrin, Hutchinson, and Hake, 1966.

21. Kelly and Hake, 1970.

22. Kupfer, Allen, and Malagodi, 2008.

23. Azrin, 1960; Miller, 1960.

24. Pryor, 1999, p. 107.

25. The potential for abuse, and for imitation, for example. Researchers have shown that kids tend to imitate behaviors that have been rewarded. Albert Bandura famously demonstrated that preschoolers who watched adults kicking and otherwise abusing a Bobo doll were more likely to be aggressive with other toys: Bandura, Ross, and Ross, 1961.

26. Azrin, 1970.

27. Roderick, Pitchford, and Miller, 1997.

28. Bloomsmith et al., 1994.

29. Timeout in various forms has been used for centuries, of course. It was formally described and tested in the animal lab in the 1950s by B. F. Skinner himself, among others. For people, a number of researchers first published on timeout in the early 1960s,

including Nathan Azrin, Montrose Wolf, and Arthur Staats. The earliest published use of the term in human application research that I've been able to locate is Wolf, Risley, and Mees, 1964.

30. Perone, 2003.

CHAPTER 7. CHOICES AND SIGNALS

1. Nestle, 2006, p. 17.
2. Quay, 1959.
3. Borrero et al., 2007.
4. Davison and McCarthy, 1988.
5. For choice and differential equations, see, for example, Machado, Keen, and Macaux, 2009.
6. Schneider and Davison, 2005. For more choices, see, for example, Rothstein, Jensen, and Neuringer, 2008. Also see Schneider, 2008.
7. Cows: Matthews and Temple, 1979. Chickens: Sumpter, Temple, and Foster, 1998. Bluegills: Banna, DeVries, and Newland, 2011. Brushtail possums: Bron et al., 2003. Coyotes: Gilbert-Norton, Shahan, and Shivik, 2009.
8. Social reinforcers: Borrero et al., 2007. Brain stimulation: Hollard and Davison, 1971. Cartoons: Epstein et al., 1991. Drugs: Anderson and Woolverton, 2000. Money: Magoon and Critchfield, 2008. Magoon and Critchfield also describe the differential-outcomes effect, in which (put simply) different consequences have different influences. Warmth: Silberberg, Thomas, and Berendzen, 1991.
9. Wild birds in their native habitats: Houston, 1986. Arithmetic problems: Neef et al., 1992.
10. Godin and Keenleyside, 1984.
11. Gray, 1994.
12. McDowell et al., 2008.
13. Sumpter, Foster, and Temple, 2002.
14. McAdie, Foster, and Temple, 1996.
15. Pedersen et al., 2005.
16. Rashotte, Foster, and Austin, 1984.
17. Rajala and Hantula, 2000.
18. Academic tasks: Mace, McCurdy, and Quigley, 1990. Methadone administration: Spiga et al., 2005.
19. Reed, Critchfield, and Martens, 2006.
20. Kollins, Newland, and Critchfield, 1997; Madden and Perone, 1999; Neuringer, Deiss, and Imig, 2000.

21. Horne and Lowe, 1993.

22. Schneider and Davison, 2006; Schneider and Morris, 1992.

23. Epstein et al., 1991.

24. Snyder and Patterson, 1995.

25. Thaler and Sunstein, 2008.

26. Tinker and Tucker, 1997.

27. Teale, 1978, p. 265.

28. Bird, 2004, p. 118.

29. Levey et al., 2009.

30. Signal-detection theory, for example, as in this classic: Green and Swets, 1966. A mathematical model combining signal-detection theory and the science of consequences was introduced in Davison and Tustin, 1978.

31. McKenzie and Day, 1971.

32. Robinson, Foster, and Bridges, 1976. We start young for the matching law, too. I was the first (with a colleague) to demonstrate matching in neonates, Schneider and Lickliter, 2010a.

33. Gould, 1985.

34. Dunne, 1995, pp. 56–61; Cody, 1974, p. 203.

35. Diamond, 1992, p. 198.

36. Abramson, 1994, p. 201.

37. Swengel, 2001.

38. Zickefoose, 2000.

39. Holz and Azrin, 1961. As described in this article, this works whether the negative is a consequence for a behavior or merely acting as an associated signal.

40. Dinsmoor, 2001.

41. Beninger, Kendall, and Vanderwolf, 1974.

42. Ator and Griffiths, 1983.

43. Abbott and Badia, 1984. This literature is mixed, however, depending on a variety of factors, and sometimes unsignaled aversives are chosen.

44. Bassett and Buchanan-Smith, 2007. For the effects of signals and predictability of aversives on humans, see, for example, Lejuez et al., 2000. The observing behaviors described in chapter 9 of this book are also relevant.

45. Findley and Brady, 1965.

46. Jwaideh, 1973.

47. List and Lucking-Reiley, 2002. As is so often the case, however, there are exceptions. For those for whom the goal is a powerful consequence, emphasizing how far there is to go can sometimes be more effective than signaling progress to date: Koo and Fishbach, 2008.

48. California county birding is described on John Sterling's website, http://www.sterlingbirds.com/county_birders.htm.

49. Skinner, 1981a.

50. Sherman, 1995, p. 69. Barnabus appeared on TV and in the *New York Times*, as Sherman relates. For another example of backward chaining—rolling a kayak!—see Alderson, 2004. Alderson uses the term "backward shaping" rather than "backward chaining."

51. Mehrabian, 1971, p. 56.

CHAPTER 8. PAVLOV AND CONSEQUENCES

1. Watanabe and Mizunami, 2007.

2. Lorenzetti et al., 2005; Ostlund and Balleine, 2007; Nargeot and Simmers, 2011; Brembs and Plend, 2008. There are a number of procedural ways to distinguish between the two processes. In "negative automaintenance," for example, Pavlovian processes are set up against consequence-based processes, in a sort of competition. Additional methods include different schedule effects and different effects of signal/elicitor magnitude.

3. Extinction refers to the loss of the "conditioned response" (such as salivation) after repeated presentation of the "conditioned eliciting stimulus" (such as the bell) without the "unconditioned stimulus" (the meat).

4. MacQueen et al., 1989.

5. Russell et al., 1984.

6. Exton et al., 2000. Also see Cohen, Moynihan, and Ader, 1994; Goebel et al., 2002.

7. Woods and Ramsay, 2000.

8. Woods et al., 1977.

9. Lekander et al., 1995.

10. Ehrman et al., 1992.

11. Siegel and Ramos, 2002.

12. Siegel and Ellsworth, 1986.

13. Wager et al., 2004.

14. Guo, Wang, and Luo, 2010.

15. Ehrman et al., 1992.

16. White, 1978.

17. Cuthbert et al., 2003.

18. Anonymous, 1970.

19. Graham and Desjardins, 1980.

20. Skinner, 1980, p. 62. The letters in our names tend to be liked and preferred over other letters, presumably in part due to classical conditioning. The effect has been found to be cross-cultural: Nuttin, 1987.

21. Skinner, 1980, p. 32.

22. Franzen, 2007, p. 39. Nineteenth-century British literary figure John Ruskin named the "pathetic fallacy."

23. Kenison, 2003, pp. ix–x.

24. Singer et al., 2004.

25. Levenson and Gottman, 1983.

26. Stackhouse, 2011. Zimmern is quoted on p. 37: "Consider this: Africans think it's weird that we eat cheese. We let milk rot and dry into little squares. It's delicious, but I know Africans who can't believe we eat it. They tell me, 'That's crazy,' while they're eating a cricket."

27. Brower, 1971, p. 350.

28. Scalera and Bavieri, 2009, p. 527. I can't vouch for their accuracy, but there are some great personal stories on AskReddit at http://www.reddit.com/r/AskReddit/comments/eh6rs/is_there_any_food_you_cant_eat_anymore_because_of/ (accessed January 1, 2012).

29. Logue, 1979.

30. Neumann, 2006.

31. Bykov, 1957, as described in Catania, 2006, p. 201.

32. Pavlov established this phenomenon with his research on "conditioning to time." See LaBarbera and Church, 1974; Lejeune and Wearden, 2006; Lockhart, 1966.

33. Other factors like consequences are important too. Classical conditioning might enhance the spring anticipation through association with characteristic environmental features such as flowering trees and warmer weather.

34. Zimmerman, 1957. For more on this way of establishing learned consequences, see, for example, one classic and one recent study: Salzinger et al., 1968; Isaksen and Holth, 2009.

35. Brake, 1981.

36. Sullivan and Leon, 1987.

37. There's some evidence that learned and unlearned consequences share some similar effects in the brain, as we'd expect: Cox, Andrade, and Johnsrude, 2005.

38. Youngentob and Glendinning, 2009; Youngentob et al., 2007.

39. Staats, 1968.

40. Olson and Fazio, 2001.

41. Staats and Staats, 1958.

42. See, for example, Field and Moore, 2005; Armel et al., 2009; also see Razran, 1955.

CHAPTER 9. OBSERVING AND ATTENDING

1. Chein and Schneider, 2005. Also see Moradi, Buracas, and Buxton, 2012; Gailliot et al., 2007.

2. Janata, Tillmann, and Bharucha, 2002.

3. Merzenich and deCharms, 1996.

4. There's even research showing that the matching law can account for attention: Hoch and Symons, 2007.

5. Watson and Sterling, 1998.

6. Kieran, 1947, p. 186. It's even possible to play a sport with a serious injury of which you're unaware.

7. Dinsmoor, Browne, and Lawrence, 1972; Fantino and Case, 1983; Fantino and Silberberg, 2010.

8. Fantino, 2008.

9. Jwaideh and Mulvaney, 1976.

10. Fantino and Silberberg, 2010.

11. Brock and Balloun, 1967.

12. Karlsson, Loewenstein, and Seppi, 2009.

13. Admiral Byrd and the stock market: Byrd, 1938. He did eventually ask, got bad news, and was sorry he had.

14. Tavris and Aronson, 2007, chapter 5.

15. Centers for Disease Control and Prevention, "HIV Prevention in the United States at a Critical Crossroads," August 2009, http://www.cdc.gov/hiv/resources/reports/hiv_prev_us.htm (accessed January 2, 2012).

16. Fest, 1999, p. 329.

17. Svartdal, 1991.

18. Svartdal, 1992.

19. Hefferline and Keenan, 1963.

20. Skinner, 1957a, p. 124.

21. Skinner, 1980, p. 15.

22. Moore, 1977, p. 32.

23. Epictetus, *Discourses*, 108 CE.

24. Bargh and Chartrand, 1999.

25. Hilty, 1994, p. 237.

26. Palameta and Lefebvre, 1985.

27. Valsecchi et al., 1994.

28. Dorrance and Zentall, 2001; also see Howard and White, 2003.

29. Safriel, Ens, and Kaiser, 1996; Wunderle, 1991. Some oystercatchers do not get the chance to observe their parents and pick up foraging techniques on their

own or through observing unrelated birds. The oystercatcher's bill changes shape in response to whatever it's eating, aiding this behavioral flexibility. (In a number of ways, experience can change form as well as behavior.) Learning through observation has been documented in many other species as well, including king rails and red-winged blackbirds.

30. Avital and Jablonka, 2000, pp. 81, 118. As Avital and Jablonka note, the "teaching parent" has been documented in a number of species, including domestic cats.

31. Nishida, 1987. Other macaque cultures have developed different traditions: Nakamichi et al., 1998.

32. Jablonka and Lamb, 2005.

33. Pruetz and Bertolani, 2007.

34. Van Schaik et al., 2003.

35. Krützen et al., 2005. A number of additional species have been documented to have cultures.

36. Holzhaider, Hunt, and Gray, 2010a; Holzhaider, Hunt, and Gray, 2010b. Relatedly, for herons' and other birds' "fishing" with bait such as floating bread, see Ruxton and Hansell, 2011. And, in a report from a respected ornithology journal, this very intriguing incident: At an Arizona bird feeder, an American crow was attacked by a Steller's jay. After failing repeatedly, the jay broke off a four-inch twig, then, "holding the twig in its bill with the pointed, narrow end outward, it landed on the platform and lunged at the crow, narrowly missing it." The crow threatened the jay, who dropped the stick and jumped backward. The crow then picked up the twig and lunged at the jay, pursuing the jay when it flew. Balda concluded this was tool use and weapon use: Balda, 2007.

37. Custance, Whiten, and Bard, 1995.

38. Poulson et al., 2002. Also see, e.g., Poulson et al., 1991; Learmonth, Lamberth, and Rovee-Collier, 2004.

39. Skinner, 1980, p. 351.

40. Erjavec and Horne, 2008.

41. Calvo-Merino et al., 2005. Relatedly, just imagining yourself doing an action also appears to activate the motor cortex. For example, see Cattaneo et al., 2009.

42. Pryor, "The Rhino Likes Violets," *Psychology Today*, April 1981. A reprint of this excellent article is also available online: http://www.reachingtheanimalmind.com/pdfs/ch_07/ch_07_pdf_03.pdf.

CHAPTER 10. THINKING AND COMMUNICATING

1. Reynolds, 1961.

2. Macario, 1991.

3. Itard, 1962 (original work published 1801 and 1806), pp. 75–76. For an application remediating undergeneralization in a teenager with autism see, for example, Walpole, Roscoe, and Dube, 2007.

4. Watanabe, Sakamoto, and Wakita, 1995.

5. Herrnstein, Loveland, and Cable, 1976; Herrnstein and de Villiers, 1980; Bhatt et al., 1988.

6. Vaughan, 1988. Similarly, "learning set," a form of pattern learning, was once considered to be beyond animals like rats, but they were later shown to be able to handle it: Slotnick, Hanford, and Hodos, 2000.

7. Katz, Wright, and Bodily, 2007.

8. Giurfa et al., 2001.

9. Murphy and Cook, 2008; Wright, Cook, and Kendrick, 1989.

10. This skill is known as *transitive inference*. Vasconcelos, 2008; Zentall et al., 2008; Wynne, 1997.

11. Schuster, 1983.

12. Pfungst, 1911 (originally published in German, 1907).

13. Summarized in Schusterman, 2008.

14. Manabe, Kawashima, and Staddon, 1995. For a good example, here's a link to a video of a parrot at the Knoxville Zoo: http://www.youtube.com/embed/nbrTOcUnjNY.

15. Schusterman, 2008, pp. 50–51. Also see Pryor, 1999, p. 153. To hear the seal that could talk, listen at http://www.st-andrews.ac.uk/~wtsf/Hoover.html (accessed August 12, 2012). More information can be found at http://www.findagrave.com/cgi-bin/fg.cgi?page=gr&GRid=25081724 (accessed August 15, 2012). Primates *can* sometimes acquire new, flexible sounds; see, for example, Wich et al., 2009.

16. Gibbons may have more vocal flexibility than had been thought: Koda et al., 2007.

17. Seyfarth and Cheney, 1986.

18. Pilley and Reid, 2011. A different dog was reportedly taught to use a keyboard to make requests: Rossi and Ades, 2008.

19. Herman, Richards, and Wolz, 1984.

20. Lyn et al., 2011; Taglialatela, Savage-Rumbaugh, and Baker, 2003. Making stone tools: Schick et al., 1999. Playing Pac-Man: as shown on the TV show "Champions of the Wild," season 4, episode 3, broadcast October 30, 2000.

21. Lieberman, 2002, pp. 136–42.

22. Skinner, 1986.

23. Skinner, 1957b, pp. 187–88.

24. Savage-Rumbaugh, Rumbaugh, and Boysen, 1978; Savage-Rumbaugh, 1984.

25. Itard, 1962/1801/1806, p. 217.

26. Lamarre and Holland, 1985.

27. Petursdottir, Carr, and Michael, 2005.

28. For example, Wallace, Iwata, and Hanley, 2006.

29. Kuczaj, 1978; Maratsos, 1983.

30. Tomasello, 2000; also see Goldberg, 2009. Tomasello found that concrete items were learned first.

31. Bialystok, 1997.

32. Routh, 1969.

33. Goldstein, King, and West, 2003; also see, for example, Poulson, 1983.

34. Gros-Louis et al., 2006. For contingent responsiveness in general, see Van Egeren, Barratt, and Roach, 2001.

35. Brigham and Sherman, 1968. In addition, just as children enjoy picking out tunes they've heard on a piano (it's intrinsically reinforcing), they may enjoy trying to repeat words they've heard.

36. Moskowitz, 1978. In an older study of more than fifty hearing children of deaf parents, most failed to develop normal speech or language use: Schiff and Ventry, 1976.

37. Moerk, 1990.

38. Ibid. Also see, for example, Whitehurst and Valdez-Menchaca, 1988.

39. Hart and Risley, 1995. Authors of one national education report stated that *Meaningful Differences in the Everyday Experience of Young American Children* should be "essential reading" in teachers' colleges: Walsh, Glaser, and Wilcox, 2006, p. 36.

40. Hart and Risley, 1995.

41. Skinner, 1963, p. 953.

42. Skinner, 1957b, pp. 130–38.

43. Hackenberg and Joker, 1994. Also see, for example, Hayes, Brownstein, and Greenway, 1986; Sloutsky and Fisher, 2008.

44. Michael, 1987, p. 39.

45. McClellan, 2010.

46. Lieberman, 2002.

47. Quoted in Bates and Dick, 2002, p. 294.

48. Bates and Roe, 2001.

49. Vargha-Khadem et al., 1997.

50. Carroll, 2005b; Li et al., 2007. Indeed, some data suggest that the fruit fly version of *FOXP2* is involved in learning from consequences.

PART 3: SHAPING DESTINIES

CHAPTER 11. EVERYDAY CONSEQUENCES

1. Serling, 1957, p. 13.
2. Gould, 2008, p. 78.
3. Cavell, 2005, p. 340.
4. Beaglehole, 1974, pp. 170–71.
5. Bryson, 1999, p. 136.
6. Gewirtz and Baer, 1958. Also see Vollmer and Iwata, 1991.
7. Hindman, "Loneliness Can Drive Elderly to Trust Telemarketers," http://www.silverplanet.com/scams/protecting-seniors/loneliness-can-drive-elderly-trust-telemarketers/6696, September 1, 2008 (accessed January 6, 2012).
8. Turnbull, 1978; Gardner, 2000.
9. Hokanson, Willers, and Koropsak, 1968.
10. Maccoby, 2003, pp. 132–35. Given anonymity, no gender differences in aggressiveness: for example, Lightdale and Prentice, 1994. In many cases, even without anonymity, there's no gender differences in aggressiveness. See Carol Tavris's classic summary: Tavris, 1989, chapter 7. More recently, see Richardson and Hammock, 2007. Similarly, Klein and Hodges (2001) confirmed other studies suggesting that gender differences in empathy are due primarily to differences in motivation. For example, payment for empathic accuracy considerably improved the scores of both men and women, with no statistically significant gender differences.
11. Hokanson, Willers, and Koropsak, 1968.
12. Milgram interviewed in Tavris, 1974, p. 72. Living with different social norms can make them seem normal. In the Peace Corps, in a patriarchal culture, being subordinate and serving men began to become automatic for many of my fellow female volunteers. It was modeled and reinforced.
13. Latané and Nida, 1981.
14. Steinbeck, 1962, p. 8.
15. Krebs, 1975.
16. Blair et al., 1997.
17. On the mixed motives for altruism, see, for example, Batson and Shaw, 1991.
18. Midlarsky and Bryan, 1967.
19. See the summary of others' work in Avital and Jablonka, 2000, pp. 182–83.
20. Thornton and McAuliffe, 2006.
21. One-day-old rats: Hoffman, Flory, and Alberts, 1999. One-day-old chicks: Delsaut, 1991.
22. Eleven studies using these choice procedures are summarized in

Granier-Deferre et al., 2011. Also see DeCasper and Spence, 1986; Darcheville, Boyer, and Miossec, 2004.

23. Brackbill, 1958; Brossard and Decarrie, 1968; Siqueland and Lipsitt, 1966.

24. Pryor, 1999, p. 63. Shaping was very fast. The longtime bestseller *Toilet Training in Less Than a Day* relies on shaping to deliver the results it describes (actually achieved by many, although often taking longer): Azrin and Foxx, 1974.

25. Phelan, 2003, p. 120.

26. Ibid., pp. 122–23; Kazdin, 2008, appendix.

27. One good source is Pryor, 1999. Shaping can work *too* well: two researchers shaped rats to stick their heads farther and farther over the edge of a platform—until they eventually fell off (onto a net)! Rasey and Iversen, 1993.

28. Savage, 2001.

29. See Karen Pryor's description of the "training game" in her book *Don't Shoot the Dog!* 1999, pp. 52–58.

30. Skinner, 1983, pp. 150–51.

31. Snyder and Patterson, 1995.

32. For example, Webster-Stratton, Reid, and Hammond, 2004; Taylor and Biglan, 1998; Brestan and Eyberg, 1998.

33. Brestan and Eyberg, 1998.

34. Faber and Mazlish, 1999, p. 187.

35. Hart et al., 1968.

36. Turnbull, 1978; Gardner, 2000. American Academy of Pediatrics child-rearing policies against corporal punishment: no author, healthychildren.org, "Where We Stand: Spanking," http://www.healthychildren.org/English/family-life/family-dynamics/communication-discipline/Pages/Where-We-Stand-Spanking.aspx (accessed January 7, 2012). Updated May 19, 2011.

Recommending timeouts: Markarian, "Positive Parenting: How to Encourage Good Behavior," *Healthy Children Magazine*, Winter 2008, 22–23. Updated May 19, 2011. Available online at http://www.healthychildren.org/English/family-life/family-dynamics/Pages/Positive-Parenting-How-To-Encourage-Good-Behavior.aspx (accessed January 7, 2012).

Many nations now ban corporal punishment of children.

37. Everett et al., 2010. There are quite a few different forms of timeout.

38. Cipani, 2004, p. 91. Available at http://www.ecipani.com/PoT.pdf.

39. Zeilberger, Sampen, and Sloane, 1968.

40. Faber and Mazlish, 1999, pp. 112–13.

41. Kazdin and Rotella, "No, You Shut Up! What to Do When Your Kid Provokes You into an Inhuman Rage" at http://www.slate.com/articles/life/family/2009/02/no_you_shut_up.single.html (accessed January 7, 2012). Updated February 5, 2009.

42. Pryor, 1999, pp. 124–25.

43. Faber and Mazlish, 1999, p. 268.

44. Gottman, 1994. Note that these are usually presumed rather than demonstrated rewards and negatives.

45. Rath and Clifton, 2004.

46. For education and business, see coverage of the ratio in chapter 14. Child-rearing: Greene et al., 1999. Teenaged boys in a group home: Friman et al., 1997. Prison rehabilitation: Gendreau and Goggin, 1996. For prison rehabilitation, also see French and Gendreau, 2006.

47. Fredrickson and Losada, 2005.

48. Sutherland, 2008.

49. Baldwin and Baldwin, 2001, pp. 217–18.

50. Baumeister et al., 2003. Also see Bronson, "How Not to Talk to Your Kids: The Inverse Power of Praise," *New York* magazine, February 19, 2007, http://nymag .com/news/features/27840/index2.html.

Modeling can be involved too, for example, as in this classic study on modeled rules about standards of achievement: Bandura and Kupers, 1964.

51. Blackwell, Trzesniewski, and Dweck, 2007.

52. Mueller and Dweck, 1998.

CHAPTER 12. FIGHTING THE IMPULSE

1. Iversen and Mogensen, 1988.

2. Skinner, 1938, p. 73.

3. Lattal and Gleeson, 1990.

4. Lattal and Metzger, 1994.

5. Okouchi, 2009.

6. Carpenter, 1986. Carpenter concluded that scurvy may have been a contributing factor in the tragic end of Scott's polar party. Certainly their diet was lacking in vitamin C.

7. Statistics on safety-belt use available from the US Centers for Disease Control and Prevention, "Adult Seat Belt Use in the US," http://www.cdc.gov/vitalsigns/Seat BeltUse/ (accessed July 21, 2011).

8. Mazur, 1996.

9. Well summarized in Ainslie, 1992. The hyperbolic curve for value starts at a high level with no delay, then drops off quickly. For a general review of the effects of delays, see Schneider, 1990. For the generality of hyperbolic delay discounting, see Green and Myerson, 2004. For an economic perspective, see Frederick, Loewenstein, and O'Donoghue, 2002. For other factors including framing and classical conditioning, see Ainslie, 2009.

10. Bertilson and Dengerink, 1975.

11. Loewenstein, 1987.

12. This observation was investigated experimentally by Ostaszewski, Green, and Myerson, 1998.

13. Rachlin and Green, 1972.

14. Jimura et al., 2009.

15. Mazur and Biondi, 2009.

16. Jackson and Hackenberg, 1996.

17. Shoda, Mischel, and Peake, 1990. Again, modeling can be influential, as in this classic study: Bandura and Mischel, 1965. Among the many factors are amount of relative delay, size, nature of the consequences and behaviors, other consequences available, history, rules, visibility, signals, sequence of events, and role models from whom learning through observation is influential.

18. Moffitt et al., 2011.

19. Mischel and Ebbesen, 1970, p. 335.

20. Grosch and Neuringer, 1981.

21. Shead and Hodgins, 2009. Also see Reynolds, 2006; Kollins, 2003.

22. Perry et al., 2005.

23. Simon, Mendez, and Setlow, 2007.

24. Schweitzer and Sulzer-Azaroff, 1988.

25. Skinner, 1948, pp. 107–14.

26. Van Haaren, Van Hest, and Van De Poll, 1988. Because of delays between each trial, delay-discounting studies like this one are set up so larger-later choices bring higher overall reinforcement rates.

27. Quote about hypermiling: "An Excellent Look at Hypermiling," http://www.priusownersgroup.com/?p=4180 (accessed January 9, 2012).

28. Dobson and Griffin, 1992; also see Karjalainen, 2011.

29. Dingfelder, 2006, p. 60.

30. Ariely and Wertenbroch, 2002.

31. Polivy and Herman, 2002.

32. Dunleavey, "How to Pay Off $4 Million of Debt," http://articles.money central.msn.com/Investing/HomeMortgageSavings/HowToPayOff4MillionOfDebt .aspx (accessed January 9, 2012), quote on birthday check from p. 1.

Dunleavey, "Feel Like You're Drowning in Debt? Get Help Online," http://www.goodhousekeeping.com/family/budget/paying-down-debt-2 (accessed January 9, 2012), quote on smileys from p. 2.

33. Typing provides more good examples of automatic chains. Many words become single units, such that I sometimes find myself starting to type one word and absentmindedly typing a more common word that begins the same way.

34. Higgins et al., 1991.

35. Schlinger, Blakeley, and Kaczor, 1990.

36. Ainslie, 1974.

37. Siegel and Rachlin, 1995.

38. Thaler and Benartzi, 2004.

39. Lyon, 2008, p. 862.

40. Trollope and Hemingway's charts: Wallace, 1977.

41. Pryor, 1999, p. 64.

42. Catherine the Great quoted in Rounding, 2006, p. 188.

43. Kjelle, 2008, p. 44. If you're flying solo, watch out for "bootleg" reinforcement—not actually living up to your goal but taking the reward anyway. You may need others to help, or you may need stronger commitments and consequences. Take advantage of the matching law and make sure another source of rewards is available. Skinner recommended working on two projects at the same time, getting the benefits of variety as well.

44. Gneezy and Rustichini, 2000.

45. Centers for Disease Control and Prevention data: "Obesity and Overweight," http://www.cdc.gov/nchs/fastats/overwt.htm (accessed January 9, 2012).

46. Lowe et al., 2004. The program has now been tested with thousands of children. More information on the Irish program is available at http://www.fooddudes .ie/main.html. For the main Food Dudes website, see "The Food Dudes Behaviour Change Programme for Healthy Eating," http://www.fooddudes.co.uk/. Societally, we could be taking other steps as well. Nutrition professor Marion Nestle (*What to Eat*, p. 522) suggests, for example, subsidizing sustainable food production and promoting healthier products.

47. Nelson, 1997, p. 160. Also see Freedman, 2011.

48. Critchfield, 1999. Recording too frequently, however, eliminated the gains.

49. De Luca and Holborn, 1992.

50. Ayres, 2010. See stickK at http://www.stickK.com/ (accessed January 9, 2012). As a personal example, Ayres likes the game Minesweeper, so he deletes it from his computer when he buys a new one. He also uses a screener so he doesn't get tempted to read too much about sports. It's embarrassing to admit, he says, but these commitment tactics enhance his productivity (p. 30).

51. Ibid.

CHAPTER 13. ENDANGERED SPECIES, UNDERCOVER CROWS, AND THE FAMILY DOG

1. A number of the Humane Society of the United States' online resources emphasize the benefits of positive-reinforcement methods: http://www.humanesociety .org (accessed January 10, 2012).

2. Statistics from "U.S. Pet Ownership Statistics," Humane Society of the United States, August 12, 2011, http://www.humanesociety.org/issues/pet_overpopulation/ facts/pet_ownership_statistics.html (accessed January 10, 2012).

3. Minta, Minta, and Hunting 1992.

4. Antle, 2011.

5. Owen, "Animal Odd Couple: New Book Documents Adventures of TK the Cat, Tonda the Orangutan," *Panama City News Herald*, April 11, 2010, http:// www.newsherald.com/articles/tonda-82956-friend-appear.html (accessed January 10, 2012).

6. Line, "One Picture: Monkeying Around," *Audubon*, March 2008, p. 160.

7. Pryor, 1999, pp. 72–73.

8. Pryor, "Charging the Clicker," August 1, 2006, http://www.clickertraining .com/node/824 (accessed January 10, 2012).

9. Pryor, 2009, p. 159.

10. Coppinger and Coppinger, 2001, pp. 202–203.

11. Shelters that discourage the use of choke, prong, or shock collars can be found in Missouri, Colorado, and Oregon, for example. The Humane Society of Boulder Valley has received a lot of publicity for its stand, at one time even sponsoring an exchange program for nonaversive harnesses: "No-Choke Challenge!" http://content.boulder humane.org/nochoke/. Also see http://www.co.washington.or.us/HHS/Animal Services/AnimalShelter/Adoption/upload/Dog-Adoption-Packet-on-Web.pdf and "Mostly Mutts Animal Rescue," http://www.mostlymutts.net/ (all accessed January 10, 2012).

12. From a story described in Toft, "My Journey to All Positive Reinforce-ment Training," 2007, http://r-plusdogtraining.info/about.htm (accessed January 10, 2012).

13. Pryor, 2009, pp. 122–23.

14. Toft, "My Journey."

15. Sutherland, 2008, p. 76. Also see Pryor, 1999, p. 174. On p. xii, Pryor notes "I stopped yelling at my kids," seeing that positive approaches worked better.

16. Reinhardt, 2003.

17. Shyne and Block, 2010.

18. Kastelein and Wiepkema, 1988.

19. Savastano, Hanson, and Savastano, 2003, p. 258.

20. Colahan and Breder, 2003, p. 235.

21. Ethier and Balsamo, 2005.

22. Muraco and Stamper, 2003.

23. Dorey et al., 2009.

24. Shepherdson et al., 1993, p. 215.

25. Maher, 2005.

26. Pryor, 1981. Available at http://www.reachingtheanimalmind.com/pdfs/ch_07/ch_07_pdf_03.pdf.

27. Maple, 2007.

28. Hanson, Larson, and Snowdon, 1976.

29. Markowitz, 1978.

30. Markowitz and Line, 1989. For environmental enrichment in general, see Tarou and Bashaw, 2007.

31. "Animal Training Philosophy at SeaWorld & Busch Gardens," http://www.seaworld.org/animal-info/info-books/training/animal-training-philosophy.htm (accessed January 10, 2012).

32. Medina et al., 2005. For lab animals, see, for example, Olsson and Dahlborn, 2002; Reese, 1991; and one of my earliest studies, referred to earlier, Schneider, 1988.

33. Mark, "Reducing Stress in Northern Bald Ibis through Training," 2007. Available online at http://reachingtheanimalmind.com/pdfs/ch_07/ch_07_pdf_02 .pdf. Also see Karen Pryor's "Playtime," http://www.clickertraining.com/node/105. The Animal Behavioral Management Alliance, in whose newsletter the Mark article appeared, is the chief zoo organization devoted to the science of consequences. Also see Desmond and Laule, 1994.

34. Moir, 2006, pp. 135–40.

35. Angulo, 2004. For a similar effort with the New Zealand robin, see Maloney and McLean, 1995.

36. In partnership with the US Army, Working Dogs for Conservation is described in Overton, "A Dog 'Tail' of Two Snails," US Army website, March 29, 2010, http://www.army.mil/article/36531/A_dog___039_tail__039__of_two _snails/ (accessed January 10, 2012).

37. Cablk and Heaton, 2006.

38. US Fish & Wildlife Service, National Wildlife Refuge System, "Turtle Dogs to the Rescue," http://www.fws.gov/refuges/mediatipsheet/August_2010/TurtleDogs totheRescue.html (accessed August 10, 2012). Updated June 19, 2012.

39. Ernst et al., 2005.

40. McAdie et al., 2005.

41. Hughes and Black, 1973.

42. Arave et al., 1984.

43. Winter and Hillerton, 1995.

44. Karen Pryor, "Sheep Dogs, Sheep, and Signals," May 15, 2007, http://www .clickertraining.com/node/1253 (accessed January 10, 2012).

45. Silva, "Sheep Training with Clicker: Target and Agility Exercices [*sic*]," training in 2006, video uploaded April 12, 2007, http://www.youtube.com/watch? v=YKnSls3zz9A (accessed January 10, 2012). Silva's website: http://www.educa-cao .blogspot.com.

46. Ferguson and Rosales-Ruiz, 2001.

47. Pryor, 2009, p. 219.

48. Clicker training for Endal: "Assistance Dogs Are Trained as Partners for the Disabled," August 10, 2000, http://articles.cnn.com/2000-08-10/health/super.dog _1_training-dogs-partons-nice-dog/2?_s=PM:HEALTH (accessed August 22, 2011). Canine Partners for Independence did the clicker training. Endal's exploits and quote: "Obituary: Endal GM," March 23, 2009, http://www.pdsa.org.uk/about-us/ media-pr-centre/news/1027_obituary-%E2%80%93 -endal-gm (accessed August 22, 2011).

49. Sonoda et al., 2011.

50. Fjellanger, Andersen, and McLean, 2002.

51. Poling et al., 2011.

52. "Working towards a Mine-free Mozambique," APOPO nonprofit organization (Dutch acronym, in English means Anti-Personnel Landmines Detection Product Development), http://www.apopo.org/cms.php?cmsid=25&lang=en (accessed January 10, 2012).

53. McLaughlin, "Giant Rats Put Noses to Work on Africa's Land Mine Epidemic," CNN, September 8, 2010, http://www.cnn.com/2010/WORLD/africa/09/07/ herorats.detect.landmines/index.html (accessed January 10, 2012).

54. Coté, *San Francisco Chronicle*, "Navy to Showcase Trained Marine Mammals in Bay," May 18, 2010, http://articles.sfgate.com/2010-05-18/bay-area/ 20902768_1 _navy-marine-mammal-program-sea-lions-diver (accessed January 11, 2012).

55. Russian World War II tank attack dogs: Zaloga, 1989, p. 43. "Anti-tank Dog," http://en.wikipedia.org/wiki/Anti-tank_dogs (January 11, 2012). Many original sources are in Russian.

56. "UK Pondered Suicide Pigeon Attacks," BBC News, May 21, 2004, http:// news.bbc.co.uk/2/hi/uk_news/3732755.stm (accessed January 11, 2012).

"War Pigeon, http://en.wikipedia.org/wiki/War_pigeon (accessed January 11, 2012).

B. F. Skinner was involved in a wartime project to train pigeons to guide missiles before reasonably accurate mechanical guidance systems had been developed: Skinner, 1960.

57. Pigeon Search and Rescue Project (PROJECT SEA HUNT), summary of US

Coast Guard report, http://www.uscg.mil/history/articles/PigeonSARProject.asp (last modified May 28, 2009, accessed January 11, 2012).

58. "Military Studied Using Trained Crows to Search for Osama bin Laden," NBC News, May 6, 2011, http://www.nbcactionnews.com/dpp/news/national/military-studied-using-trained-crows-to-search-for-osama-bin-laden#ixzz1jAeX9k4Y (accessed January 11, 2012).

59. Crows: Marzluff et al., 2010. In this case, masks were used on a number of different people, a "caveman" face versus Dick Cheney's. Pigeons: Belguermi et al., 2011. In this study, no masks were used. Discriminating between two women of similar age and skin color provided a serious challenge for the wild pigeons.

60. Teale, 1948, p. 89.

CHAPTER 14. THE REWARDS OF EDUCATION AND WORK

1. Rosenthal and Jacobson, 1968; expanded edition, 1992.

2. Harris and Rosenthal, 1985.

3. Esquith, 2003, p. 176.

4. Escalante and Dirmann, 1990, p. 411.

5. Ibid., p. 414.

6. Ibid, p. 409.

7. Walker, 1979, p. 168; Thomas, Becker, and Armstrong, 1968.

8. O'Leary et al., 1970. For additional research showing this effect, see Chance, 2008, p. 133. Note also that the process of correcting a mistake need not be punishing: for example, it can be done in the spirit of "You're making good progress, and here's an opportunity to learn."

9. Madsen, Becker, and Thomas, 1968.

10. Kazdin, 1973.

11. Tingstrom, Sterling-Turner, and Wilczynski, 2006.

12. Saigh and Umar, 1983.

13. Embry, 2002.

14. Tanol et al., 2010.

15. Esquith's system: Esquith, 2003.

16. Canter, 2010.

17. Posted by Marcia on June 5, 2002. Suggestions from many teachers: "The Behavior Management Page," http://www.teachingheart.net/classroombehaviormanage.html (accessed January 11, 2012).

18. Several studies are summarized in Chance, 2008, pp. 17–18.

19. Latham, 1992.

20. Mattaini, 2001, p. 59.

21. Madsen et al., 1970. Also see Mayer et al., 1993. Considering the alternative choices and their rewards is also helpful, and again, these competing schedules and behaviors can be quantified as in chapter 7: Billington and DiTommaso, 2003. Finally, see Reschly, 2008, on "best practices" in school psychology.

22. Glynn, Thomas, and Shee, 1973.

23. Walker, Mattson, and Buckley, 1971.

24. Walker, 1979, p. 291. The benefits of surprise rewards are echoed in Chance, 2008, p. 112.

25. Devers, Bradley-Johnson, and Johnson, 1994.

26. Duckworth et al., 2011.

27. Brinch and Galloway, 2012. Also see Williams, 1998.

28. Classic study: Flynn, 1987. Recent update, including Kenya: Daley et al., 2003.

29. Harrell, Woodyard, and Gates, 1955.

30. For example, Campbell et al., 2002.

31. Ericsson, 1993; Bloom, 1985.

32. Chase and Simon, "Perception in Chess."

33. Feltovich, Prietula, and Ericsson, 2006. Not surprisingly, Skinner also recognized the role of schedules of reinforcement in developing perseverance: Skinner, 1968, pp. 165–66.

34. Bloom, 1985, p. 514.

35. Shenk, 2010, p. 43.

36. Gladwell, 2008, p. 268

37. Watkins, 1997; Borman et al., 2003.

38. Ibid. The US Office of Education funded the worst-performing programs because they were more consistent with existing educational philosophy.

39. Dobbie, Fryer, and Fryer, 2011, p. 158. Some questions remain.

40. Angrist et al., 2010. Also see Henig, 2008.

41. Halpin and Halpin, 1982.

42. Fryer, 2010.

43. Pryor, "On My Mind: Paying Kids to Learn," April 26, 2010, http://www.clickertraining.com/node/2857 (accessed January 11, 2012).

44. Cuban, "Paying Students to Do Well in School: What Economists Are Learning about Pay-4-Performance," April 16, 2010, http://larrycuban.wordpress.com/2010/04/16/paying-students-to-do-well-in-school-what-economists-are-learning-about-pay-4-performance/ (accessed January 11, 2012).

Willingham, "An Analysis of Pay-for-Grades Schemes," May 19, 2010, http://voices.washingtonpost.com/answer-sheet/daniel-willingham/the-psychology-behind-paying-k.html (accessed January 11, 2012).

As a sample of research on the cost-benefit analyses of these methods, see Blonigen et al., 2008; Putnam et al., 2002.

45. Ripley, "Should Kids Be Bribed to Do Well in School?" *Time*, April 5, 2010, http://www.time.com/time/nation/article/0,8599,1978589,00.html (accessed January 11, 2012).

46. Skinner, 1968.

47. Chance, 2008, chapter 8 provides a good summary. We've already seen Fryer's support.

48. Morgan, 1984, p. 9.

49. Cameron, Banko, and Pierce, 2001. Also see Cameron and Pierce, 2002.

50. Levitt and Dubner, 2005, p. 13.

51. For more on the benefits of control, see, for example, Hockey and Earle, 2006.

52. Murphy, 1947, p. 21.

53. Franzen, 2002.

54. Other consequences were also present, as described in Parsons, 1978.

55. Daniels, "A Reversal of Fortunes: Who Is Really Appraised by the Performance Appraisal Process?" June 29, 2011, http://aubreydanielsblog.com/page/2/ (accessed January 11, 2012).

56. Blanchard and Johnson, 1982, p. 97.

57. Bales, 1950.

58. Rath and Clifton, 2004.

59. Losada and Heaphy, 2004.

60. Also note football-helmet stickers for good plays, widely used in the NCAA.

61. Daniels, "Parenting and Behavior: Examples from Real Parents," August 30, 2011, http://aubreydanielsblog.com/2011/08/30/parenting-and-behavior-examples-from-real-parents/ (accessed January 11, 2012).

62. Summarized in Bucklin and Dickinson, 2001.

63. Hughes, 1986.

64. Hogan, Bell, and Olson, 2009, p. 15.

65. Pedalino and Gamboa, 1974.

66. Camden, Price, and Ludwig, 2011.

67. Daniels, 2000.

68. Johnson and Dickinson, 2010.

69. Kortick and O'Brien, 1996.

70. Bateman and Ludwig, 2003.

71. Sulzer-Azaroff and Austin, 2000. Also see Krause, Seymour, and Sloat, 1999.

72. Alavosius and Sulzer-Azaroff, 1986.

73. Monaco, Olsson, and Hentges, 2005, p. 621.

74. Ludwig and Geller, 2000.

75. Fox, Hopkins, and Anger, 1987.

76. Woolfolk, Castellan, and Brooks, 1983.

77. Till, Stanley, and Priluck, 2008.

78. Louie, Kulik, and Jacobson, 2001.

79. Pratkanis and Aronson, 1991.

80. Prelec and Simester, 2001.

81. Kivetz, Urminsky, and Zheng, 2006.

82. Schlosser, 2001.

CHAPTER 15. HELP FOR ADDICTION, AUTISM, AND OTHER CONDITIONS

1. Stressful life events: Kessler, 1997. Childhood trauma: Heim and Nemeroff, 2001. Also see Stroud et al., 2011.

2. Orwell, 1950.

3. Leshan and Worthington, 1956.

4. Leight and Ellis, 1981.

5. Humphrey and Krout, 1975.

6. The *Science* study: House, Landis, and Umberson, 1988. Loneliness and depression: Cornwell and Waite, 2009.

7. Myers, 1997, p. 178. Pavlovian extinction is an additional Pavlovian process involved in the therapeutic weakening of impact.

8. Pennebaker, 1997.

9. Pryor, 1987.

10. Dweck and Reppucci, 1973.

11. Hiroto, 1974.

12. Fish: Padilla et al., 1970. Roaches: Brown and Stroup, 1988. How long the effect lasts depends on a number of factors.

13. Seligman and Maier, 1967.

14. Nation and Massad, 1978.

15. Howard et al., 1986.

16. Quoted in Boswell, 1953, p. 33.

17. Ekers, Richards, and Gilbody, 2008. Also see Mazzucchelli, Kane, and Rees, 2009.

18. Hopko et al., 2003.

19. Babyak et al., 2000.

20. Daley, 2008.

21. Wolitzky-Taylor et al., 2008. For more on classical conditioning and learning from consequences in anxiety disorders, see Lohr, Olatunji, and Sawchuk, 2007.

22. Hoffman et al., 2003.

23. Pogrebin, 1996, pp. 163–65.

24. This outstanding review includes the human dog-phobia example, the dental-research example, and phobias in animals: Mineka and Zinbarg, 2006.

25. To predict the reinforcing value of drugs, scientists utilize different schedules of reinforcement as well as many other aspects of the science of consequences: Ator and Griffiths, 2003.

26. Covington and Miczek, 2001.

27. Smoking statistics can be found at the US Centers for Disease Control and Prevention, "Tobacco-Related Mortality," http://www.cdc.gov/tobacco/data_statistics/fact_sheets/health_effects/tobacco_related_mortality/index.htm (accessed January 14, 2012).

28. Forestell and Mennella, 2005.

29. The United States is due to join these nations soon.

30. Dallery, Meredith, and Glenn, 2008.

31. Volpp et al., 2009.

32. Lamb et al., 2010.

33. Roll, Higgins, and Badger, 1996.

34. In addition to specific cues, the context in general can become classically conditioned: Conklin, 2006.

35. Wilkinson and Pickett, 2009, pp. 70–71.

36. Office of National Drug Control Policy, Washington, DC, December 2004, "The Economic Costs of Drug Abuse in the United States, 1992–2002," p. vi. Available online at https://www.ncjrs.gov/ondcppubs/publications/pdf/economic_costs.pdf, accessed January 15, 2012. The figure given is for 2002.

37. Stitzer and Vandrey, 2008. For self-control methods applied to addiction, see Monterosso and Ainslie, 2009.

38. Dutra et al., 2008. Also see Prendergast et al., 2006, which mentions the employment-opportunity version of contingency management.

39. Petry et al., 2005.

40. Lovaas, 1987.

41. Rogers and Vismara, 2008.

42. US Department of Health and Human Services, 1999; Myers and Johnson, 2007, available online at http://aappolicy.aappublications.org/cgi/content/full/pediatrics;120/5/1162 (accessed January 15, 2012). On this website, "A statement of reaffirmation for this policy was published on [December 1, 2010]."

43. Young et al., 1994. As is commonly found, generalization worked much better within each category than across categories. For example, imitation training for toy play generalized well across toys but less well to pantomime movements.

44. Maurice, 1993, pp. 246–47.

45. Ibid., p. 206.

46. Iwata et al., 1994. For more on the role of attention, see Thompson and Iwata, 2001. For a positives-only way of handling challenging behaviors in the developmentally disabled in the classroom, see Davis, Fredrick, and Alberto, 2012.

47. Rincover and Devany, 1982, pp. 67–81.

48. Kahng, Iwata, and Lewin, 2002.

49. Charman et al., 1997; Dube et al., 2004; Mundy et al., 1986. Animals can also handle joint attention: Udell, Dorey, and Wynne, 2008.

50. Whalen and Schreibman, 2003; Taylor and Hoch, 2008; Klein et al., 2009; Isaksen and Holth, 2009.

51. Maurice, 1993, pp. 111–12.

52. "Attention-Deficit/Hyperactivity Disorder (ADHD): Data & Statistics," Centers for Disease Control and Prevention, http://www.cdc.gov/ncbddd/adhd/data. html, 9.5 percent incidence of ADHD as reported by parents. Updated December 12, 2011 (accessed January 15, 2012).

53. Aase and Sagvolden, 2006.

54. Antrop et al., 2006; Schweitzer and Sulzer-Azaroff, 1995.

55. Foster, 2010.

56. DeSantis, Webb, and Noar, 2008.

57. *FAA Safety Briefing*, Federal Aviation Administration newsletter, March/April 2010, Dr. Warren S. Silberman, medical certification Q&A column, http://www.faa.gov/news/safety_briefing/2010/media/MarApr2010.pdf (accessed January 15, 2012).

58. Ayllon, Layman, and Kandel, 1975.

59. Molina et al., 2009.

60. Pelham, 1999.

61. Ibid.

62. Fabiano et al., 2009.

63. Flor et al., 2002; Blair et al., 1997.

64. Hare, 1999.

65. Gao et al., 2010.

66. Boehm et al., 1995.

67. Livingston et al., 2005.

68. Omnibus Budget Reconciliation Act of 1987.

69. Goldiamond, 1973.

70. Clare and Jones, 2008.

71. "Pediatric food refusal" is the technical term for this condition, which occurs most often in infants and toddlers. Prematurity and low birth weight are risk factors. Reviews: Sharp et al., 2010; Ahearn et al., 1996.

72. Piacentini et al., 2010.

CHAPTER 16. CONSEQUENCES ON A GRAND SCALE

1. "My Lai Pilot Hugh Thompson," transcript of a report from National Public Radio, January 6, 2006, http://www.npr.org/templates/story/story.php?storyId=5133444 (accessed January 16, 2012).

2. Milgram, 1974. Even now, fifty years later, reading Milgram's book is still worthwhile.

3. Ibid.

4. Blass, 1999.

5. Milgram, 1974, p. 54. None of us knows what we're capable of until we're actually put to the test. When the whaling ship *Essex* was sunk by a sperm whale in 1820, the sailors struggled at sea for months in small boats, slowly starving and dying. Finally, the crew of one boat agreed to voluntary cannibalism, they drew lots, and one man was duly shot and eaten. While this makes for painful reading, most of us can understand how it could come to seem necessary, even altruistic (the chosen victim did not complain).

6. Hughes, 1986, p. 376.

7. Wilder, 1990. For an intriguing study that combines these effects with the effects of rewards, see Bettencourt et al., 1992.

8. Gilbert, Tafarodi, and Malone, 1993. Similarly, with respect to stereotypes, a classic article: Devine, 1989.

9. Told that hard work would pay off, Carol Dweck's schoolchildren did better (see chapter 11). Similarly, researchers showed the benefits of simply letting African Americans, women, and other stereotyped groups know about this self-fulfilling effect and about how flexible our abilities really are. In one well-known study, women did as well as men on a difficult math test when told that no gender differences were expected. These sorts of results have now been repeated many times. For African Americans and academic success: Aronson, Fried, and Good, 2002. For women and math success: Johns, Schmader, and Martens, 2005.

10. Including the quotes from Elliott's students (original spelling and grammar maintained): transcript of PBS *Frontline* episode, "A Class Divided," written by William Peters and Charlie Cobb, March 26, 1985, http://www.pbs.org/wgbh/pages/frontline/shows/divided/etc/script.html#ixzz 1d3nPFTJc (accessed January 16, 2012).

11. Stewart et al., 2003.

12. Sherif and Sherif, 1965, "Ingroup and Intergroup Relations," http://www.brocku.ca/MeadProject/Sherif/Sherif_1965f.html accessed January 16, 2012.

13. Shapira and Madsen, 1969.

14. Aronson and Bridgeman, 1979.

15. Singh, 1991.

16. Diener and Seligman, 2002; Putnam, 2000; Diener and Biswas-Diener, 2002.

17. Small, Loewenstein, and Slovic, 2007.

18. Eliot, 1954, p. 111 (original letter from 1859).

19. Vollaro, 2009. Available online at: http://www.historycooperative.org/journals/jala/30.1/vollaro.html.

20. Jeffrey M. Jones, "Record-High 86% Approve of Black-White Marriages," Gallup, September 12, 2011, http://www.gallup.com/poll/149390/record-high-approve-black-white-marriages.aspx (accessed January 16, 2012).

21. Jones, 2006. Consequences come on schedules in these political processes too. Consider the way taxes are seldom raised right before an election. Representatives who vote too often against their parties get punished by receiving less financial support for reelection. More bills are typically passed toward the end of a legislative session, in part because of time-based schedules: Representatives want to have something to show the public by the deadline. Some poorly considered laws get sneaked in because everyone's in a hurry to finish so they can leave.

22. Hughes, 1986, pp. 145–48. For a classic account of short-term versus long-term consequences for individuals and businesses, see Veblen, 1994. Also note that colonial governments sometimes taxed the natives to force them to work. Otherwise, they didn't need money. For example, a "hut tax" was imposed by the British in South Africa, in part to raise a workforce for infrastructure projects.

23. Perrin, 1979. Another example of conflicting consequences: Tobacco advertising brings in much-needed immediate cash to developing nations even as it blights the longer-term health of their citizens: Stebbins, 2001.

24. Bjornlund and Bjornlund, 2010.

25. Asimov, 1972, p. 237.

26. Jackson et al., 2001.

27. Bright, 1998; van Driesche and van Driesche, 2000.

28. Hager, 2006.

29. Diamond, 2005.

30. Ostrom, 2007. Also see Ostrom, 2009.

31. Gerrodette and Rojas-Bracho, 2011.

32. Balmford et al., 2002.

33. Adams et al., 2004.

34. Balmford et al., 2002. Also see Balmford et al., 2011.

35. Skinner, 1983, p. 223.

36. Diamond, 2005, p. 485.

37. Allcott, 2011. One recent study on water conservation found that the reference to social norms was an essential part of the motivation: Ferraro, Miranda, and

Price, 2011. For an excellent review of energy-conservation studies, see Abrahamse et al., 2005.

38. Emily Green, "DWP Offers Cash Incentive to Remove Lawns," *Los Angeles Times*, June 13, 2009, http://articles.latimes.com/2009/jun/13/home/hm-grass13 (accessed January 17, 2012).

39. Robert Kunzig, "Drying of the West," *National Geographic*, February 2008, pp. 90–113. Online at http://www.exloco.org/dwnld/Natl%20Geo%20Drying%20 of%20the%20West.pdf (accessed January 17, 2012).

40. Kohlenberg and Phillips, 1973. Another study on littering looked at rules, "us" versus "them" social norms, and direct consequences: Miller, Brickman, and Bolen, 1975.

41. Lanphear et al., 2005.

42. Reyes, 2007.

43. Many laboratory studies showed that animals of a variety of species were sensitive to small amounts of lead exposure—as measured by changes in their behavior on the fixed-interval schedule of reinforcement. For example, Cory-Slechta, Pokora, and Preston, 1996.

44. Goldstein, Cialdini, and Griskevicius, 2008.

45. Jablonka and Lamb, 2005, p. 74.

46. Indeed, of the 7 billion people now on the planet, most live at material levels well below those in the industrialized nations—but that's another book.

47. Heckman and Masterov, 2007; Heckman, 2007.

BIBLIOGRAPHY

Aase, H., and T. Sagvolden. "Infrequent, but Not Frequent, Reinforcers Produce More Variable Responding and Deficient Sustained Attention in Young Children with Attention-Deficit/Hyperactivity Disorder (ADHD)." *Journal of Child Psychology and Psychiatry* 47 (2006): 457–71.

Abbey, E. *Desert Solitaire.* New York: McGraw-Hill, 1968.

Abbott, B. B., and P. Badia. "Preference for Signaled over Unsignaled Shock Schedules: Ruling Out Asymmetry and Response Fixation as Factors." *Journal of the Experimental Analysis of Behavior* 41 (1984): 45–52.

Abrahamse, W., L. Steg, C. Vlek, and T. Rothengatter. "A Review of Intervention Studies Aimed at Household Energy Conservation." *Journal of Environmental Psychology* 25 (2005): 273–91.

Abramson, C. I. *A Primer of Invertebrate Learning: The Behavioral Perspective.* Washington, DC: American Psychological Association, 1994.

———. "Aversive Conditioning in Honeybees (*Apis mellifera*)." *Journal of Comparative Psychology* 100 (1986): 108–16.

Adams, W. M., R. Aveling, D. Brockington, B. Dickson, J. Elliott, J. Hutton, D. Roe, B. Vira, and W. Wolmer. "Biodiversity Conservation and the Eradication of Poverty." *Science* 306 (2004): 1146–49.

Addolorato, G., L. Leggio, A. Ferrulli, S. Cardone, L. Vonghia, A. Mirijello, L. Abenavoli, et al. "Effectiveness and Safety of Baclofen for Maintenance of Alcohol Abstinence in Alcohol-Dependent Patients with Liver Cirrhosis: Randomised, Double-Blind Controlled Study." *Lancet* 370 (2007): 1915–22.

Aharon, I., N. Etcoff, D. Ariely, C. F. Chabris, E. O'Connor, and H. C. Breiter. "Beautiful Faces Have Variable Reward Value: fMRI and Behavioral Evidence." *Neuron* 32 (2001): 537–51.

Ahearn, W. H., M. E. Kerwin, P. Eicher, J. Shantz, and W. Swearingin. "An Alternating Treatments Comparison of Two Intensive Interventions for Food Refusal." *Journal of Applied Behavior Analysis* 29 (1996): 321–32.

Ainslie, G. "Hyperbolic Discounting versus Conditioning and Framing as the Core Process in Addictions and Other Impulses." In *What Is Addiction?* edited by D. Ross, H. Kincaid, D. Spurrett, and P. Collins, pp. 211–45. Cambridge, MA: MIT Press, 2009.

Ainslie, G. "Impulse Control in Pigeons." *Journal of the Experimental Analysis of Behavior* 21 (1974): 485–89.

Ainslie, G. *Picoeconomics: The Strategic Interaction of Successive Motivational States within the Person*. Cambridge: Cambridge University Press, 1992.

Alavosius, M. P., and B. Sulzer-Azaroff. "The Effects of Performance Feedback on the Safety of Client Lifting and Transfer." *Journal of Applied Behavior Analysis* 19 (1986): 261–67.

Alderson, D. "Back into Rolling." *Sea Kayaker* (December 2004): 22–27.

Allcott, H. "Social Norms and Energy Conservation." *Journal of Public Economics* 95 (2011): 1082–95.

Amedi, A., L. B. Merabet, J. Camprodon, F. Bermpohl, S. Fox, I. Ronen, D. Kim, and A. Pascual-Leone. "Neural and Behavioral Correlates of Drawing in an Early Blind Painter: A Case Study." *Brain Research* 1242 (2008): 252–62.

Anderson, K. G., and W. L. Woolverton. "Concurrent Variable-Interval Drug Self-Administration and the Generalized Matching Law: A Drug-Class Comparison." *Behavioural Pharmacology* 11 (2000): 413–20.

Anderson, R. C. "Operant Conditioning and Copulation Solicitation Display Assays Reveal a Stable Preference for Local Song by Female Swamp Sparrows *Melospiza georgiana*." *Behavioral Ecology and Sociobiology* 64 (2009): 215–23.

Angrist, J. D., S. M. Dynarski, T. J. Kane, P. A. Pathak, and C. R. Walters. "Who Benefits from KIPP?" National Bureau of Economic Research Working Paper 15740, February 2010.

Angulo, P. F. "Dispersion, Supervivencia y Reproduccion de la Pava Aliblanca *Penelope albipennis Taczanowski* 1877 (*Cracidae*) Reintroducida a su Habitat Natural en Peru" [Dispersal, Survival and Reproduction of Reintroduced White-Winged Guan *Penelope albipennis Taczanowski*, 1877 (*Cracidae*) to Its Natural Habitat]. In Spanish with an English summary. *Ecologia Aplicada* 3 (2004): 2112–17.

Anokhin, K. V., and S. P. R. Rose. "Learning-Induced Increase of Immediate Early Gene Messenger RNA in the Chick Forebrain." *European Journal of Neuroscience* 3 (1991): 162–67.

Anonymous. "Effects of Sexual Activity on Beard Growth in Men." *Nature* 226 (1970): 869–70.

Antle, B. *Suryia and Roscoe: The True Story of an Unlikely Friendship*. New York: Holt, 2011.

Antrop, I., P. Stock, S. Verté, J. R. Wiersema, D. Baeyens, and H. Roeyers. "ADHD and Delay Aversion: The Influence of Non-Temporal Stimulation on Choice for Delayed Rewards." *Journal of Child Psychology and Psychiatry* 47 (2006): 1152–58.

Arave, C. W., W. Temple, J. V. Leman, and R. Kilgour. "Discriminability and Preference among Milking Machine Functions by Dairy Cows." *Applied Animal Behavior Science* 12 (1984): 313–25.

Ariely, D., and K. Wertenbroch. "Procrastination, Deadlines, and Performance: Self-Control by Precommitment." *Psychological Science* 13 (2002): 219–24.

Armel, K. C., C. Pulido, J. T. Wixted, and A. A. Chiba. "The Smart Gut: Tracking Affective Associative Learning with Measures of 'Liking,' Facial Electromyography, and Preferential Looking." *Learning and Motivation* 40 (2009): 74–93.

Aronson, E., and D. Bridgeman. "Jigsaw Groups and the Desegregated Classroom: In Pursuit of Common Goals." *Personality and Social Psychology Bulletin* 5 (1979): 438–46.

Aronson, J., C. B. Fried, and C. Good. "Reducing the Effects of Stereotype Threat on African American College Students by Shaping Theories of Intelligence." *Journal of Experimental Social Psychology* 38 (2002): 113–25.

Asimov, I. *The Gods Themselves*. Greenwich, CT: Fawcett Crest, 1972.

Ator, N. A., and R. R. Griffiths. "Lorazepam and Pentobarbital Drug Discrimination in Baboons: Cross-Drug Generalization and Interaction with Ro 15-1788." *Journal of Pharmacology and Experimental Therapeutics* 226 (1983): 776–82.

———. "Principles of Drug Abuse Liability Assessment in Laboratory Animals." *Drug and Alcohol Dependence* 70 (2003): S55–S72.

Avital, E., and E. Jablonka. *Animal Traditions: Behavioural Inheritance in Evolution*. Cambridge: Cambridge University Press, 2000.

Ayllon, T., and N. H. Azrin. "Punishment as a Discriminative Stimulus and Conditioned Reinforcer with Humans." *Journal of the Experimental Analysis of Behavior* 9 (1966): 411–19.

Ayllon, T., D. Layman, and H. J. Kandel. "A Behavioral-Educational Alternative to Drug Control of Hyperactive Children." *Journal of Applied Behavior Analysis* 8 (1975): 137–46.

Ayres, I. *Carrots and Sticks: Unlock the Power of Incentives to Get Things Done*. New York, Bantam, 2010.

Azrin, N. H. "Effects of Punishment Intensity during Variable-Interval Reinforcement." *Journal of the Experimental Analysis of Behavior* 3 (1960): 123–42.

———. "Punishment of Elicited Aggression." *Journal of the Experimental Analysis of Behavior* 14 (1970): 7–10.

———. "Time-Out from Positive Reinforcement." *Science* 133 (1961): 382–83.

Azrin, N. H., and R. M. Foxx. *Toilet Training in Less Than a Day: A Tested Method for Teaching Your Child Quickly and Easily*. New York: Simon and Schuster, 1974.

Azrin, N. H., R. R. Hutchinson, and D. F. Hake. "Extinction-Induced Aggression." *Journal of the Experimental Analysis of Behavior* 9 (1966): 191–204.

Babyak, M., J. A. Blumenthal, S. Herman, P. Khatri, M. Doraiswamy, K. Moore, W. E. Craighead, T. T. Baldewicz, and K. R. Krishnan. "Exercise Treatment for Major Depression: Maintenance of Therapeutic Benefit at 10 Months." *Psychosomatic Medicine* 62 (2000): 633–38.

Bacon, W. E., and I. G. Wong. "Reinforcement Value of Electrical Brain Stimulation in Neonatal Dogs." *Developmental Psychobiology* 5 (1972): 195–200.

Balaban, P. M., and R. Chase. "Self-Stimulation in Snails." *Neuroscience Research Communications* 4 (1989): 139–47.

Balda, R. P. "Corvids in Combat: With a Weapon?" *Wilson Journal of Ornithology* 119 (2007): 100–102.

Baldwin, J. D., and J. L. Baldwin. *Behavior Principles in Everyday Life.* 4th ed. Upper Saddle River, NJ: Prentice-Hall, 2001.

Bales, R. F. *Interaction Process Analysis: A Method for the Study of Small Groups.* Chicago: University of Chicago Press, 1950.

Balmford, A., A. Bruner, P. Cooper, R. Costanza, S. Farber, R. E. Green, M. Jenkins, et al. "Economic Reasons for Conserving Wild Nature." *Science* 297 (2002): 950–53.

Balmford, A., B. Fisher, R. E. Green, R. Naidoo, B. Strassburg, R. K. Turner, and A. S. L. Rodrigues. "Bringing Ecosystem Services into the Real World: An Operational Framework for Assessing the Economic Consequences of Losing Wild Nature." *Environmental and Resource Economics* 48 (2011): 161–75.

Bandura, A., and C. J. Kupers. "Transmission of Patterns of Self-Reinforcement through Modeling." *Journal of Abnormal and Social Psychology* 69 (1964): 1–9.

Bandura, A., and W. Mischel. "Modifications of Self-Imposed Delay of Reward through Exposure to Live and Symbolic Models." *Journal of Personality and Social Psychology* 2 (1965): 698–705.

Bandura, A., D. Ross, and S. A. Ross. "Transmission of Aggression through Imitation of Aggressive Models." *Journal of Abnormal and Social Psychology* 63 (1961): 575–82.

Banna, K. M., D. DeVries, and M. C. Newland "Choice in the Bluegill (*Lepomis macrochirus*)." *Behavioural Processes* 88 (2011): 33–43.

Bargh, J. A., and T. L. Chartrand. "The Unbearable Automaticity of Being." *American Psychologist* 54 (1999): 462–79.

Barnes, G. W., and A. Baron. "Stimulus Complexity and Sensory Reinforcement." *Journal of Comparative and Physiological Psychology* 54 (1961): 466–69.

Baron, A. "Avoidance and Punishment." In *Techniques in the Behavioral and Neural Sciences: Experimental Analysis of Behavior, Part 1*, edited by I. H. Iversen and K. A. Lattal, pp. 173–217. Amsterdam: Elsevier, 1991.

Bassett, L., and H. M. Buchanan-Smith. "Effects of Predictability on the Welfare of Captive Animals." *Applied Animal Behaviour Science* 102 (2007): 223–45.

Bateman, M. J., and T. D. Ludwig. "Managing Distribution Quality through an Adapted Incentive Program with Tiered Goals and Feedback." *Journal of Organizational Behavior Management* 24 (2003): 33–55.

Bates, E., and F. Dick. "Language, Gesture, and the Developing Brain." *Developmental Psychobiology* 40 (2002): 293–310.

Bates, E., and K. Roe. "Language Development in Children with Unilateral Brain Injury." In *Handbook of Developmental Cognitive Neuroscience*, edited by C. A. Nelson and M. Luciana, pp. 2–23. Cambridge, MA: MIT Press, 2001.

Bateson, P. "The Active Role of Behaviour in Evolution." In *Evolutionary Processes and Metaphors*, edited by M. W. Ho and S. W. Fox, pp. 191–207. Chichester, UK: Wiley, 1988.

Bateson, P., and E. P. Reese. "The Reinforcing Properties of Conspicuous Stimuli in the Imprinting Situation." *Animal Behaviour* 17 (1969): 692–99.

Batson, C. D., and L. L. Shaw. "Evidence for Altruism: Toward a Pluralism of Prosocial Motives." *Psychological Inquiry* 2 (1991): 107–22.

Baumeister, R. F., J. D. Campbell, J. I. Krueger, and K. D. Vohs. "Does High Self-Esteem Cause Better Performance, Interpersonal Success, Happiness, or Healthier Lifestyles?" *Psychological Science in the Public Interest* 4 (2003): 1–44.

Beaglehole, J. C. *The Life of Captain James Cook*. Stanford, CA: Stanford University Press, 1974.

Behrens, T. E. J., L. T. Hunt, M. W. Woolrich, and M. F. S. Rushworth. "Associative Learning of Social Value." *Nature* 456 (2008): 245–49.

Beisson, J., and T. M. Sonneborn. "Cytoplasmic Inheritance of the Organization of the Cell Cortex in *Paramecium aurelia*." *Proceedings of the National Academy of Sciences* 53 (1965): 275–82.

Belguermi, A., D. Bovet, A. Pascal, A. Prévot-Julliard, M. Saint Jalme, L. Rat-Fischer, and G. Leboucher. "Pigeons Discriminate between Human Feeders." *Animal Cognition* 14 (2011): 909–14.

Beninger, R. J., S. B. Kendall, and C. H. Vanderwolf. "The Ability of Rats to Discriminate Their Own Behaviours." *Canadian Journal of Psychology* 28 (1974): 79–91.

Berridge, K. C., and T. E. Robinson. "What Is the Role of Dopamine in Reward: Hedonic Impact, Reward Learning, or Incentive Salience?" *Brain Research Reviews* 28 (1998): 309–69.

Bertilson, H. S., and H. A. Dengerink. "The Effects of Active Choice, Shock Duration, Shock Experience, and Probability on the Choice between Immediate and Delayed Shock." *Journal of Research in Personality* 9 (1975): 97–112.

Bettencourt, B. A., M. B. Brewer, M. R. Croak, and N. Miller. "Cooperation and the Reduction of Intergroup Bias: The Role of Reward Structure and Social Orientation." *Journal of Experimental Social Psychology* 28 (1992): 301–19.

Bhatt, R. S., E. A. Wasserman, W. F. Reynolds, and K. S. Knauss. "Conceptual Behavior in Pigeons: Categorization of Both Familiar and Novel Examples from Four Classes of Natural and Artificial Stimuli." *Journal of Experimental Psychology: Animal Behavior Processes* 14 (1988): 219–34.

Bialystok, E. "The Structure of Age: In Search of Barriers to Second Language Acquisition." *Second Language Research* 13 (1997): 116–37.

Billington, E., and N. M. DiTommaso. "Demonstrations and Applications of the Matching Law in Education." *Journal of Behavioral Education* 12 (2003): 91–104.

Bird, D. Watching Bird Behavior, *Bird Watcher's Digest* 27, January 2004, p. 118.

Bjornlund, V., and H. Bjornlund. "Sustainable Irrigation: A Historical Perspective." In *Incentives and Instruments for Sustainable Irrigation*, edited by H. Bjornlund, pp. 13–24. Boston: WIT Press, 2010.

Blackwell, L. S., K. H. Trzesniewski, and C. S. Dweck. "Implicit Theories of Intelligence Predict Achievement Across an Adolescent Transition: A Longitudinal Study and an Intervention." *Child Development* 78 (2007): 246–63.

Blair, R. J. R., L. Jones, F. Clark, and M. Smith. "The Psychopathic Individual: A Lack of Responsiveness to Distress Cues?" *Psychophysiology* 34 (1997): 192–98.

Blanchard, K., and S. Johnson. *One Minute Manager*. New York: Morrow, 1982.

Blass, T. "The Milgram Paradigm after 35 Years: Some Things We Now Know about Obedience to Authority." *Journal of Applied Social Psychology* 29 (1999): 955–78.

Blonigen, B., W. Harbaugh, L. Singell, R. Horner, K. Irvin, and K. Smokowski. "Application of Economic Analysis to School-Wide Positive Behavior Support Programs." *Journal of Positive Behavior Interventions* 10 (2008): 5–19.

Blood, A. J., and R. J. Zatorre. "Intensely Pleasurable Responses to Music Correlate with Activity in Brain Regions Implicated in Reward and Emotion." *Proceedings of the National Academy of Sciences* 98 (2001): 11818–23.

Bloom, B. *Developing Talent in Young People*. New York: Ballantine, 1985.

Bloomsmith, M. A., G. E. Laule, P. L. Alford, and R. H. Thurston. "Using Training to Moderate Chimpanzee Aggression during Feeding." *Zoo Biology* 13 (1994): 557–66.

Boehm, S., A. P. Thurnau, A. L. Whall, K. L. Cosgrove, J. D. Locke, and E. A. Schlenk. "Behavioral Analysis and Nursing Interventions for Reducing Disruptive Behaviors of Patients with Dementia." *Applied Nursing Research* 8 (1995): 118–22.

Borman, G. D., G. M. Hewes, L. T. Overman, and S. Brown. "Comprehensive School Reform and Achievement: A Meta-Analysis." *Review of Educational Research* 73 (2003): 125–230.

Borrero, J. C., S. S. Crisolo, Q. Tu, W. A. Rieland, N. A. Ross, M. T. Francisco, and K. Y. Yamamoto. "An Application of the Matching Law to Social Dynamics." *Journal of Applied Behavior Analysis* 40 (2007): 589–601.

Boswell, J. *Boswell on the Grand Tour, Germany and Switzerland, 1764*. Edited by F. A. Pottle. New York: McGraw-Hill, 1953.

Brackbill, Y. "Extinction of the Smiling Response in Infants as a Function of Reinforcement Schedule." *Child Development* 29 (1958): 115–24.

Brake, S. C. "Suckling Infant Rats Learn a Preference for a Novel Olfactory Stimulus Paired with Milk Delivery." *Science* 211 (1981): 506–508.

Branch, M. N. "Behavioral Pharmacology." In *Techniques in the Behavioral and Neural Sciences: Experimental Analysis of Behavior, Part 2*, edited by I. H. Iversen and K. A. Lattal, pp. 21–77. Amsterdam: Elsevier, 1991.

Breiter, H. C., A. Itzhak, D. Kahneman, A. Dale, and P. Shizgal, "Functional Imaging of Neural Responses to Expectancy and Experience of Monetary Gains and Losses." *Neuron* 30 (2001): 619–39.

Brembs, B., and M. Heisenberg. "The Operant and the Classical in Conditioned Orientation of *Drosophila melanogaster* at the Flight Simulator." *Learning and Memory* 7 (2000): 104–15.

Brembs, B., and W. Plend. "Double Dissociation of PKC and AC Manipulations on Operant and Classical Learning in Drosophila." *Current Biology* 18 (2008): 1168–71.

Brennan, W. M., E. W. Ames, and R. W. Moore. "Age Differences in Infants' Attention to Patterns of Different Complexities." *Science* 151 (1966): 354–56.

Brestan, E. V., and S. M. Eyberg. "Effective Psychosocial Treatments of Conduct-Disordered Children and Adolescents: 29 Years, 82 Studies, and 5,272 Kids." *Journal of Clinical Child Psychology* 27 (1998): 180–89.

Brigham, T. A., and J. A. Sherman. "An Experimental Analysis of Verbal Imitation in Preschool Children." *Journal of Applied Behavior Analysis* 1 (1968): 151–58.

Bright, C. *Life Out of Bounds: Bioinvasion in a Borderless World*. New York: Norton, 1998.

Brinch, C. N., and T. A. Galloway. "Schooling in Adolescence Raises IQ Scores." *Proceedings of the National Academy of Sciences* 109 (2012): 425–30.

Brock, T. C., and J. L. Balloun. "Behavioral Receptivity to Dissonant Information." *Journal of Personality and Social Psychology* 6 (1967): 413–28.

Bron, A., C. E Sumpter, T. M. Foster, and W. Temple. "Contingency Discriminability, Matching, and Bias in the Concurrent-Schedule Responding of Possums (*Trichosurus vulpecula*)." *Journal of the Experimental Analysis of Behavior* 79 (2003): 289–306.

Brossard, L. M., and Decarrie, T. G. "Comparative Reinforcing Effect of Eight Stimulations on the Smiling Response of Infants." *Journal of Child Psychology and Psychiatry* 9 (1968): 51–59.

Brower, L. P. "Prey Coloration and Predator Behavior." In *Topics in Animal Behavior, Topics in the Study of Life: The BIO Source Book*, edited by V. Dethier. New York: Harper and Row, 1971. Quoted in E. Fantino and C. A. Logan, *The Experimental Analysis of Behavior: A Biological Perspective*. San Francisco: Freeman, 1979.

Brown, G. E., and K. Stroup. "Learned Helplessness in the Cockroach (*Periplaneta americana*)." *Behavioral and Neural Biology* 50 (1988): 246–50.

Brumberg, J. S., and F. H. Guenther. "Development of Speech Prostheses: Current Status and Recent Advances." *Expert Review of Medical Devices* 7 (2010): 667–79.

Bryson, B. *I'm a Stranger Here Myself: Notes on Returning to America after Twenty Years Away*. New York: Random House, 1999.

Bucklin, B. R., and A. M. Dickinson. "Individual Monetary Incentives: A Review of

Different Types of Arrangements between Performance and Pay." *Journal of Organizational Behavior Management* 21 (2001): 45–137.

Butler, R. A. "Incentive Conditions which Influence Visual Exploration." *Journal of Experimental Psychology* 48 (1954): 19–23.

Bykov, K. M. *The Cerebral Cortex and the Internal Organs.* Translated by W. H. Gantt. New York: Chemical Publishing, 1957.

Byrd, R. E. *Alone.* New York: Putnam, 1938.

Cablk, M. E., and J. S. Heaton. "Accuracy and Reliability of Dogs in Surveying for Desert Tortoise (*Gopherus agassizii*)." *Ecological Applications* 16 (2006): 1926–35.

Calvo-Merino B., D. E. Glaser, J. Grèzes, R. E. Passingham, and P. Haggard. "Action Observation and Acquired Motor Skills: An fMRI Study with Expert Dancers." *Cerebral Cortex* 15 (2005): 1243–49.

Camden, M. C., V. A. Price, and T. D. Ludwig. "Reducing Absenteeism and Rescheduling among Grocery Store Employees with Point-Contingent Rewards." *Journal of Organizational Behavior Management* 31 (2011): 140–49.

Cameron, J., K. M. Banko, and W. D. Pierce. "Pervasive Negative Effects of Rewards on Intrinsic Motivation: The Myth Continues." *Behavior Analyst* 24 (2001): 1–44.

Cameron, J., and W. D. Pierce. *Rewards and Intrinsic Motivation: Resolving the Controversy.* Westport, CT: Bergin and Garvey, 2002.

Campbell, F. A., C. T. Ramey, E. Pungello, J. Sparling, and S. Miller-Johnson. "Early Childhood Education: Young Adult Outcomes from the Abecedarian Project." *Applied Developmental Science* 6 (2002): 42–57.

Canter, L. *Assertive Discipline: Positive Behavior Management for Today's Classroom.* 4th ed. Bloomington, IN: Solution Tree, 2010.

Carlstead, K. "Effects of Captivity on the Behavior of Wild Mammals." In *Wild Mammals in Captivity: Principles and Techniques*, edited by D. G. Kleiman, M. E. Allen, K. V. Thompson, and S. Lumpkin, pp. 317–33. Chicago: University of Chicago Press, 1996.

Carlstead, K., J. Seidensticker, and R. Baldwin. "Environmental Enrichment for Zoo Bears." *Zoo Biology* 10 (1991): 3–16.

Carmena J. M., M. A. Lebedev, R. E. Crist, J. E. O'Doherty, D. M. Santucci, F. D. Dragan, P. G. Patil, C. S. Henriquez, and M. A. L. Nicolelis. "Learning to Control a Brain-Machine Interface for Reaching and Grasping by Primates." *PLoS Biology* 1 (2003): e42.

Carpenter, K. J. *The History of Scurvy and Vitamin C.* Cambridge: Cambridge University Press, 1986.

Carroll, S. B. *Endless Forms Most Beautiful: The New Science of Evo Devo and the Making of the Animal Kingdom.* New York: W. W. Norton, 2005a.

Carroll, S. B. "Evolution at Two Levels: On Genes and Form." *PLoS Biology* 3 (2005b): e245.

Catania, A. C. *Learning*. 4th ed. Cornwall-on-Hudson, NY: Sloan, 2006.

Catania, A. C., and Sagvolden, T. "Preference for Free Choice over Forced Choice in Pigeons." *Journal of the Experimental Analysis of Behavior* 34 (1980): 77–86.

Cattaneo, L., C. Fausto, A. Jezzini, and G. Rizzolatti. "Representation of Goal and Movements without Overt Motor Behavior in the Human Motor Cortex: A Transcranial Magnetic Stimulation Study." *Journal of Neuroscience* 29 (2009): 11134–38.

Cavell, S. "The Good of Film." In *Cavell on Film*, edited by W. Rothman, pp. 333–48. Albany: State University of New York Press, 2005.

Cerutti, D., and A. C. Catania. "Pigeons' Preference for Free Choice: Number of Keys versus Key Area." *Journal of the Experimental Analysis of Behavior* 68 (1997): 349–56.

Chance, P. *Learning and Behavior*. 5th ed. Belmont, CA: Wadsworth, 2003.

———. *The Teacher's Craft: The 10 Essential Skills of Effective Teaching*. Long Grove, IL: Waveland Press, 2008.

Chapin, J. K., K. A. Moxon, R. S. Markowitz, and M. A. L. Nicolelis. "Real-Time Control of a Robot Arm Using Simultaneously Recorded Neurons in the Motor Cortex." *Nature Neuroscience* 2 (1999): 664–70.

Charman, T., J. Swettenham, S. Baron-Cohen, A. Cox, G. Baird, and A. Drew. "Infants with Autism: An Investigation of Empathy, Pretend Play, Joint Attention, and Imitation." *Developmental Psychology* 33 (1997): 781–89.

Chase, W. G., and H. A. Simon. "Perception in Chess." *Cognitive Psychology* 4 (1973): 55–81.

Chein, J. M., and W. Schneider. "Neuroimaging Studies of Practice-Related Change: fMRI and Meta-Analytic Evidence of a Domain-General Control Network for Learning." *Cognitive Brain Research* 25 (2005): 607–23.

Cherek, D. R., and R. Pickens. "Schedule-Induced Aggression as a Function of Fixed-Ratio Value." *Journal of the Experimental Analysis of Behavior* 14 (1970): 309–11.

Cierpial, M., and R. McCarty. "Hypertension in SHR Rats: Contribution of Maternal Environment." *American Journal of Physiology: Heart and Circulatory Physiology* 253 (1987): 980–84.

Cipani, E. *Punishment on Trial: A Resource Guide to Child Discipline*. Reno, NV: Context Press, 2004.

Clare, L., and R. S. P. Jones. "Errorless Learning in the Rehabilitation of Memory Impairment: A Critical Review." *Neuropsychology Review* 18 (2008): 1–23.

Cody, M. *Competition and the Structure of Bird Communities*. Princeton, NJ: Princeton University Press, 1974.

Cohen, N., J. A. Moynihan, and R. Ader. "Pavlovian Conditioning of the Immune System." *International Archives of Allergy and Immunology* 105 (1994): 101–106.

Colahan, H., and C. Breder. "Primate Training at Disney's Animal Kingdom." *Journal of Applied Animal Welfare Science* 6 (2003): 235–46.

Collias, E. C., and N. E. Collias. "The Development of Nest-Building Behavior in a Weaverbird." *Auk* 81 (1964): 42–52.

Conklin, C. A. "Environments as Cues to Smoke: Implications for Human Extinction-Based Research and Treatment." *Experimental and Clinical Psychopharmacology* 14 (2006): 12–19.

Connor, J. *The Complete Birder: A Guide to Better Birding*. New York: Houghton Mifflin, 1988.

Cook, L., and A. C. Catania. "Effects of Drugs on Avoidance and Escape Behavior." *Federation Proceedings* 23 (1964): 818–35.

Coppinger, R., and L. Coppinger. *Dogs: A Startling New Understanding of Canine Origin, Behavior, and Evolution*. New York: Scribner, 2001.

Cornwell, E. Y., and L. J. Waite. "Social Disconnectedness, Perceived Isolation, and Health among Older Adults." *Journal of Health and Social Behavior* 50 (2009): 31–48.

Cory-Slechta, D. A., M. J. Pokora, and R. A. Preston. "The Effects of Dopamine Agonists on Fixed Interval Schedule-Controlled Behavior Are Selectively Altered by Low-Level Lead Exposure." *Neurotoxicology and Teratology* 18 (1996): 565–75.

Cory-Slechta, D. A., M. B. Virgolini, A. Rossi-George, D. Weston, and M. Thiruchelvam. "Experimental Manipulations Blunt Time-Induced Changes in Brain Monoamine Levels and Completely Reverse Stress, but Not Pb+/– Stress-Related Modifications to These Trajectories." *Behavioural Brain Research* 205 (2009): 76–87.

Covington, H. E., and K. A. Miczek. "Repeated Social-Defeat Stress, Cocaine or Morphine: Effects on Behavioral Sensitization and Intravenous Cocaine Self-Administration 'Binges.'" *Psychopharmacology* 158 (2001): 388–98.

Cox, S. M. L., A. Andrade, and I. S. Johnsrude. "Learning to Like: A Role for Human Orbitofrontal Cortex in Conditioned Reward." *Journal of Neuroscience* 25 (2005): 2733–40.

Critchfield, T. S. "An Unexpected Effect of Recording Frequency in Reactive Self-Monitoring." *Journal of Applied Behavior Analysis* 32 (1999): 389–91.

Cross, H. A., C. G. Halcomb, and W. W. Matter. "Imprinting or Exposure Learning in Rats Given Early Auditory Stimulation." *Psychonomic Science* 7 (1967): 233–34.

Custance, D. M., A. Whiten, and K. A. Bard. "Can Young Chimpanzees (*Pan troglodytes*) Imitate Arbitrary Actions? Hayes and Hayes (1952) Revisited." *Behaviour* 132 (1995): 837–59.

Cuthbert, B. N., P. J. Lang, C. Strauss, D. Drobes, C. J. Patrick, and M. M. Bradley. "The Psychophysiology of Anxiety Disorder: Fear Memory Imagery." *Psychophysiology* 40 (2003): 407–22.

Daley, A. "Exercise and Depression: A Review of Reviews." *Journal of Clinical Psychology in Medical Settings* 15 (2008): 140–47.

Daley, T. C., S. E. Whaley, M. D. Sigman, M. P. Espinosa, and C. Neumann. "IQ on the Rise: The Flynn Effect in Rural Kenyan Children." *Psychological Science* 14 (2003): 215–19.

Dallery, J., S. Meredith, and I. M. Glenn. "A Deposit Contract Method to Deliver Abstinence Reinforcement for Cigarette Smoking." *Journal of Applied Behavior Analysis* 41 (2008): 609–15.

Daniels, A. C. *Bringing Out the Best in People: How to Apply the Astonishing Power of Positive Reinforcement.* Revised ed. New York: McGraw-Hill, 2000.

Darcheville, J. C., C. Boyer, and Y. Miossec. "Training Infant Reaching Using Mother's Voice as Reinforcer." *European Journal of Behavior Analysis* 5 (2004): 43–51.

Davis, D. H., L. D. Fredrick, and P. A. Alberto. "Functional Communication Training without Extinction Using Concurrent Schedules of Differing Magnitudes of Reinforcement in Classrooms." *Journal of Positive Behavior Interventions* 14 (2012): 162–72.

Davison, M., and D. McCarthy. *The Matching Law: A Research Review.* Hillsdale, NJ: Erlbaum, 1988.

Davison, M., and R. D. Tustin. "The Relation between the Generalized Matching Law and Signal-Detection Theory." *Journal of the Experimental Analysis of Behavior* 29 (1978): 331–36.

DeCasper, A. J., and M. J. Spence. "Prenatal Maternal Speech Influences Newborns' Perception of Speech." *Infant Behavior and Development* 9 (1986): 133–50.

Decker, S. H. "Collective and Normative Features of Gang Violence." *Justice Quarterly* 13 (1996): 243–64.

DeLange, R. J., D. M. Fambrough, E. L. Smith, and J. Bonner. "Calf and Pea Histone IV. II. The Complete Amino Acid Sequence of Calf Thymus Histone IV: Presence of {varepsilon}-N-Acetyllysine." *Journal of Biological Chemistry* 244 (1969): 319–34.

Delsaut, M. "Influence of Nonobvious Learning on the Development of the Approach Response in Chicks (*Gallus gallus*)." *International Journal of Comparative Psychology* 4 (1991): 239–51.

De Luca, R. V., and S. W. Holborn. "Effects of a Variable-Ratio Reinforcement Schedule with Changing Criteria on Exercise in Obese and Nonobese Boys." *Journal of Applied Behavior Analysis* 25 (1992): 671–79.

Department of Health and Human Services. *Mental Health: A Report of the Surgeon General.* Rockville, MD: Department of Health and Human Services, Substance Abuse and Mental Health Services Administration, Center for Mental Health Services, National Institutes of Health, National Institute of Mental Health, 1999.

DeSantis, A. D., E. M. Webb, and S. M. Noar. "Illicit Use of Prescription ADHD Medications on a College Campus: A Multimethodological Approach." *Journal of American College Health* 57 (2008): 315–23.

Desmond, T., and G. Laule. "Use of Positive Reinforcement Training in the Management of Species for Reproduction." *Zoo Biology* 13 (1994): 471–77.

Deutsch, J. A., and C. I. Howarth. "Some Tests of a Theory of Intracranial Self-Stimulation." *Psychological Review* 70 (1963): 444–60.

Devers, R., S. Bradley-Johnson, and C. M. Johnson. "The Effect of Token Reinforcement on WISC-R Performance for Fifth- through Ninth-Grade American Indians." *Psychological Record,* 44 (1994): 441–49.

Devine, P. G. "Stereotypes and Prejudice: Their Automatic and Controlled Components." *Journal of Personality and Social Psychology* 56 (1989): 5–18.

Dews, P. B. "Studies on Behavior. I. Differential Sensitivity to Pentobarbital of Pecking Performance in Pigeons Depending on the Schedule of Reward." *Journal of Pharmacology and Experimental Therapeutics* 113 (1955): 393–401.

Diamond, J. *Collapse: How Societies Choose to Fail or Succeed.* New York: Viking, 2005.

———. *The Third Chimpanzee: The Evolution and Future of the Human Animal.* New York: HarperCollins, 1992.

Diener, E., and R. Biswas-Diener. "Will Money Increase Subjective Well-Being?" *Social Indicators Research* 57 (2002): 119–69.

Diener, E., R. E. Lucas, and C. N. Scollon. "Beyond the Hedonic Treadmill: Revising the Adaptation Theory of Well-Being." *American Psychologist* 61 (2006): 305–14.

Diener, E., and M. E. P. Seligman. "Very Happy People." *Psychological Science* 13 (2002): 81–84.

Dingfelder, S. F. "Ditch the Delay Tactics: Setting Daily Goals while Keeping Broad Aims in Mind Helps Students Beat Dissertation Procrastination." *Monitor on Psychology* 37 (July 2006): 58, 60.

Dinsmoor, J. A. "Stimuli Inevitably Generated by Behavior That Avoids Electric Shock Are Inherently Reinforcing." *Journal of the Experimental Analysis of Behavior* 75 (2001): 311–33.

Dinsmoor, J. A., M. P. Browne, and C. E. Lawrence. "A Test of the Negative Discriminative Stimulus as a Reinforcer of Observing." *Journal of the Experimental Analysis of Behavior* 18 (1972): 79–85.

Dobbie, W., R. G. Fryer, and G. Fryer. "Are High-Quality Schools Enough to Increase Achievement among the Poor? Evidence from the Harlem Children's Zone." *American Economic Journal: Applied Economics* 3 (2011): 158–87.

Dobson, J. K., and J. D. A. Griffin. "Conservation Effect of Immediate Electricity Cost Feedback on Residential Consumption Behavior." *Proceedings of the 7th ACEEE Summer Study on Energy Efficiency in Buildings.* Washington, DC: American Council for an Energy-Efficient Economy, 1992.

Dodd, M. L., K. J. Klos, J. H. Bower, Y. E. Geda, K. A. Josephs, and J. E. Ahlskog. "Pathological Gambling Caused by Drugs Used to Treat Parkinson Disease." *Archives of Neurology* 62 (2005): 1377–81.

Doidge, N. *The Brain That Changes Itself: Stories of Personal Triumph from the Frontiers of Brain Science.* New York: Penguin, 2007.

Dorey, N. R., J. Rosales-Ruiz, R. Smith, and B. Lovelace. "Functional Analysis and Treatment of Self-Injury in a Captive Olive Baboon." *Journal of Applied Behavior Analysis* 42 (2009): 785–94.

Dorrance, B. R., and T. R. Zentall. "Imitative Learning in Japanese Quail (*Coturnix japonica*) Depends on the Motivational State of the Observer Quail at the Time of Observation." *Journal of Comparative Psychology* 115 (2001): 62–67.

Dowling, J. L., D. A. Luther, and P. P. Marra. "Comparative Effects of Urban Development and Anthropogenic Noise on Bird Songs." *Behavioral Ecology* 23 (2012): 201–209.

Dube, W. V., R. P. F. MacDonald, R. C. Mansfield, W. L. Holcomb, and W. H. Ahearn. "Toward a Behavioral Analysis of Joint Attention." *Behavior Analyst* 27 (2004): 197–207.

Duckworth, A. L., P. D. Quinn, D. R. Lynam, R. Loeber, and M. Stouthamer-Loeber. "Role of Test Motivation in Intelligence Testing." *Proceedings of the National Academy of Sciences* 108 (2011): 7716–20.

Dunlap, G., M. dePerczel, S. Clarke, D. Wilson, S. Wright, R. White, and A. Gomez. "Choice Making to Promote Adaptive Behavior for Students with Emotional and Behavioral Challenges." *Journal of Applied Behavior Analysis* 27 (1994): 505–18.

Dunne, P. *The Wind Masters: The Lives of North American Birds of Prey.* Boston: Houghton Mifflin, 1995.

Durrell, G. *My Family and Other Animals.* London: Penguin, 1956.

Dutra, L., G. Stathopoulou, S. L. Basden, T. M. Leyro, M. B. Powers, and M. W. Otto. "A Meta-Analytic Review of Psychosocial Interventions for Substance Use Disorders." *American Journal of Psychiatry* 165 (2008): 179–87.

Dweck, C. S., and N. D. Reppucci. "Learned Helplessness and Reinforcement Responsibility in Children." *Journal of Personality and Social Psychology* 25 (1973): 109–16.

Ehrman, R., J. Ternes, C. P. O'Brien, and A. T. McLellan. "Conditioned Tolerance in Human Opiate Addicts." *Psychopharmacology* 108 (1992): 218–24.

Eisenberger, N. I., M. D. Lieberman, and K. D. Williams. "Does Rejection Hurt? An fMRI Study of Social Exclusion." *Science* 302 (2003): 290–92.

Ekers, D., D. Richards, and S. Gilbody. "A Meta-Analysis of Randomized Trials of Behavioural Treatment of Depression." *Psychological Medicine* 38 (2008): 611–23.

Eliot, G. Vol. 3 of *The George Eliot Letters.* New Haven, CT: Yale University Press, 1954.

Embry, D. D. "The Good Behavior Game: A Best Practice Candidate as a Universal Behavioral Vaccine." *Clinical Child and Family Psychology Review* 5 (2002): 273–97.

Epstein, L. H., J. A. Smith, L. S. Vara, and J. S. Rodefer. "Behavioral Economic Analysis of Activity Choice in Obese Children." *Health Psychology* 10 (1991): 311–16.

Ericsson, K. A. "The Role of Deliberate Practice in the Acquisition of Expert Performance." *Psychological Review* 100 (1993): 363–406.

Erjavec, M., and P. J. Horne. "Determinants of Imitation of Hand-to-Body Gestures in 2- and 3-Year-Old Children." *Journal of the Experimental Analysis of Behavior* 89 (2008): 183–207.

Ernst, K., B. Puppe, P. C. Schön, and G. Manteuffel. "A Complex Automatic Feeding System for Pigs Aimed to Induce Successful Behavioural Coping by Cognitive Adaptation." *Applied Animal Behaviour Science* 91 (2005): 205–18.

Escalante, J., and J. Dirmann. "The Jaime Escalante Math Program." *Journal of Negro Education* 59 (1990): 407–23.

Esquith, R. *There Are No Shortcuts*. New York: Pantheon, 2003.

Ethier, N., and C. Balsamo. "Training Sea Turtles for Husbandry and Enrichment." In *Proceedings of the Seventh International Conference on Environmental Enrichment*, edited by N. Clum, S. Silver, and P. Thomas, pp. 106–11. New York: Shape of Enrichment, 2005.

Everett, G. E., S. D. A. Hupp, and D. J. Olmi. "Time-Out with Parents: A Descriptive Analysis of 30 Years of Research." *Education and Treatment of Children* 33 (2010): 235–59.

Exton, M. S., A. K. von Auer, A. Buske-Kirschbaum, U. Stockhorst, U. Gobel, and M. Schedlowski. "Pavlovian Conditioning of Immune Function: Animal Investigation and the Challenge of Human Application." *Behavioural Brain Research* 110 (2000): 129–41.

Faber, A., and E. Mazlish. *How to Talk So Kids Will Listen and Listen So Kids Will Talk*. New York: Harper, 1999.

Fabiano, G. A., W. E. Pelham, E. K. Coles, E. M. Gnagy, A. Chronis-Tuscanoc, and B. C. O'Connor. "A Meta-Analysis of Behavioral Treatments for Attention-Deficit/Hyperactivity Disorder." *Clinical Psychology Review* 29 (2009): 129–40.

Fantino, E. "Choice, Conditioned Reinforcement, and the Prius Effect." *Behavior Analyst* 31 (2008): 95–111.

Fantino, E., and D. A. Case. "Human Observing: Maintained by Stimuli Correlated with Reinforcement but Not Extinction." *Journal of the Experimental Analysis of Behavior* 40 (1983): 193–210.

Fantino, E., and A. Silberberg. "Revisiting the Role of Bad News in Maintaining

Human Observing Behavior." *Journal of the Experimental Analysis of Behavior* 93 (2010): 157–70.

Feltovich, P. J., M. J. Prietula, and K. A. Ericsson. "Studies of Expertise from Psychological Perspectives." In *The Cambridge Handbook of Expertise and Expert Performance*, edited by K. A. Ericsson, N. Charness, P. J. Feltovich, and R. R. Hoffman, pp. 41–67. New York: Cambridge University Press, 2006.

Ferguson, D. L., and J. Rosales-Ruiz. "Loading the Problem Loader: The Effects of Target Training and Shaping on Trailer-Loading Behavior of Horses." *Journal of Applied Behavior Analysis* 34 (2001): 409–24.

Ferraro, P. J., J. J. Miranda, and M. K. Price. "The Persistence of Treatment Effects with Norm-Based Policy Instruments: Evidence from a Randomized Environmental Policy Experiment." *American Economic Review* 101 (2011): 318–22.

Ferster, C. B., and B. F. Skinner. *Schedules of Reinforcement*. Englewood Cliffs, NJ: Prentice-Hall, 1957.

Fest, J. *Speer: The Final Verdict*. Translated by E. Osers and A. Dring. New York: Harcourt, 1999.

Fetterman, J. G., P. R. Killeen, and S. Hall. "Watching the Clock." *Behavioural Processes* 44 (1998): 211–24.

Field, A. P., and A. C. Moore. "Dissociating the Effects of Attention and Contingency Awareness on the Evaluative Conditioning Effects in the Visual Paradigm." *Cognition and Emotion* 19 (2005): 217–43.

Findley, J. D., and J. V. Brady. "Facilitation of Large Ratio Performance by Use of Conditioned Reinforcement." *Journal of the Experimental Analysis of Behavior* 8 (1965): 125–29.

Fjellanger, R., E. K. Andersen, and I. G. A. McLean. "Training Program for Filter-Search Mine Detection Dogs." *International Journal of Comparative Psychology* 15 (2002): 278–87.

Flor, H., N. Birbaumer, C. Hermann, S. Ziegler, and C. J. Patrick. "Aversive Pavlovian Conditioning in Psychopaths: Peripheral and Central Correlates." *Psychophysiology* 39 (2002): 505–18.

Flynn, J. R. "Massive IQ Gains in 14 Nations: What IQ Tests Really Measure." *Psychological Bulletin* 101 (1987): 171–91.

Forestell, C. A., and J. A. Mennella. "Children's Hedonic Judgments of Cigarette Smoke Odor: Effects of Parental Smoking and Maternal Mood." *Psychology of Addictive Behaviors* 19 (2005): 423–32.

Foster, S. "Impaired Behavior Regulation under Conditions of Concurrent Variable Schedules of Reinforcement in Children with ADHD." *Journal of Attention Disorders* 13 (2010): 358–68.

Fox, D. K., B. L. Hopkins, and W. K. Anger. "The Long-Term Effects of a Token

Economy on Safety Performance in Open-Pit Mining." *Journal of Applied Behavior Analysis* 20 (1987): 215–24.

Fraga, M. F., E. Ballestar, M. F. Paz, S. Ropero, F. Setien, M. L. Ballestar, D. Heine-Suñer et al. "Epigenetic Differences Arise during the Lifetime of Monozygotic Twins." *Proceedings of the National Academy of Sciences* 102 (2005): 10604–10609.

Francis, D. D., J. Diorio, P. M. Plotsky, and M. J. Meaney. "Enrichment and Environmental Enrichment Reverses the Effects of Maternal Separation on Stress Reactivity." *Journal of Neuroscience* 22 (2002): 7840–43.

Franzen, J. *The Discomfort Zone*. New York: Picador, 2007.

———. "Lost in the Mail." Reprinted in *How to Be Alone: Essays*. New York: Farrar, Straus, and Giroux, 2002.

Frederick, S., G. Loewenstein, and T. O'Donoghue. "Time Discounting and Time Preference: A Critical Review." *Journal of Economic Literature* 40 (2002): 351–401.

Fredrickson, B. L., and M. F. Losada. "Positive Affect and the Complex Dynamics of Human Flourishing." *American Psychologist* 60 (2005): 678–86.

Freedman, D. H. "How to Fix the Obesity Crisis." *Scientific American* 304 (February 2, 2011): 20–27.

French, S. A., and P. Gendreau. "Reducing Prison Misconducts: What Works!" *Criminal Justice and Behavior* 33 (2006): 185–218.

Friman, P. C., M. Jones, G. Smith, D. L. Daly, and R. Larzelere. "Decreasing Disruptive Behavior by Adolescent Boys in Residential Care by Increasing Their Positive to Negative Interactional Ratios." *Behavior Modification* 21 (1997): 470–86.

Fryer, R. G. "Financial Incentives and Student Achievement: Evidence from Randomized Trials." National Bureau of Economic Research Working Paper 15898, 2010.

Gailliot, M. T., R. F. Baumeister, C. N. DeWall, J. K. Maner, E. A. Plant, D. M. Tice, L. E. Brewer, and B. J. Schmeichel. "Self-Control Relies on Glucose as a Limited Energy Source: Willpower Is More Than a Metaphor." *Journal of Personality and Social Psychology* 92 (2007): 325–36.

Gao, Y., A. Raine, P. H. Venables, M. E. Dawson, and S. A. Mednick. "Association of Poor Childhood Fear Conditioning and Adult Crime." *American Journal of Psychiatry* 167 (2010): 56–60.

Gardner, P. M. "Respect and Nonviolence among Recently Sedentary Paliyan Foragers." *Journal of the Royal Anthropological Institute* 6 (2000): 215–36.

Gendreau, P., and C. Goggin. "Principles of Effective Correctional Programming." *Forum on Corrections Research* 8 (1996): 38–41.

Gentry, W. D. "Fixed-Ratio Schedule-Induced Aggression." *Journal of the Experimental Analysis of Behavior* 11 (1968): 813–17.

Gerrodette, T., and L. Rojas-Bracho. "Estimating the Success of Protected Areas for the Vaquita *Phocoena sinus*." *Marine Mammal Science* 27 (2011): E101–E125.

Gewirtz, J. L., and D. M. Baer. "Deprivation and Satiation of Social Reinforcers as Drive Conditions." *Journal of Abnormal and Social Psychology* 57 (1958): 165–72.

Gilbert, D. T., R. W. Tafarodi, and P. S. Malone. "You Can't Not Believe Everything You Read." *Journal of Personality and Social Psychology* 65 (1993): 221–33.

Gilbert, S. F., and E. M. Jorgensen. "Wormwholes: A Commentary on K. F. Schaffner's 'Genes, Behavior and Developmental Emergentism.'" *Philosophy of Science* 65 (1998): 259–66.

Gilbert-Norton, L. B., T. A. Shahan, and J. A. Shivik. "Coyotes (*Canis latrans*) and the Matching Law." *Behavioural Processes* 82 (2009): 178–83.

Ginsburg, S., and E. Jablonka. "The Evolution of Associative Learning: A Factor in the Cambrian Explosion." *Journal of Theoretical Biology* 266 (2010): 11–20.

Giurfa, M., S. Zhang, A. Jenett, R. Menzel, and M. V. Srinivasan. "The Concepts of 'Sameness' and 'Difference' in an Insect." *Nature* 410 (2001): 930–33.

Gladwell, M. *Outliers: The Story of Success*. New York: Little, Brown, 2008.

Glaser, R., W. P. Lafuse, R. H. Bonneau, C. Atkinson, and J. K. Kiecolt-Glaser. "Stress-Associated Modulation of Proto-Oncogene Expression in Human Peripheral Blood Leukocytes." *Behavioral Neuroscience* 107 (1993): 525–29.

Glynn, E. L., J. D. Thomas, and S. M. Shee. "Behavioral Self-Control of On-Task Behavior in an Elementary Classroom." *Journal of Applied Behavior Analysis* 6 (1973): 105–13.

Gneezy, U., and A. Rustichini. "A Fine Is a Price." *Journal of Legal Studies* 29 (2000): 1–17.

Godin, J. G. J., and M. H. A. Keenleyside. "Foraging on Patchily Distributed Prey by a Cichlid Fish (*Teleosti, Cichlidae*): A Test of the Ideal Free Distribution Theory." *Animal Behaviour* 32 (1984): 120–31.

Goebel, M. U., A. E. Trebst, J. Steiner, Y. F. Xie, M. S. Exton, S. Frede, A. Canbay, M. C. Michel, U. Heemann, and M. Schedlowski. "Behavioral Conditioning of Immunosuppression Is Possible in Humans." *FASEB Journal* 16 (2002):1869–73.

Goetz, E. M., and D. M. Baer. "Social Control of Form Diversity and the Emergence of New Forms in Children's Blockbuilding." *Journal of Applied Behavior Analysis* 6 (1973): 209–17.

Goldberg, A. E. "The Nature of Generalization in Language." *Cognitive Linguistics* 20 (2009): 93–127.

Goldiamond, I. "A Diary of Self-Modification." *Psychology Today* 7 (November 1973): 95–102.

Goldman, L., G. D. Coover, and S. Levine. "Bidirectional Effects of Reinforcement Shifts on Pituitary Adrenal Activity." *Physiology & Behavior* 10 (1973): 209–14.

Goldstein, M. H., A. P. King, and M. J. West. "Social Interaction Shapes Babbling: Testing Parallels between Birdsong and Speech." *Proceedings of the National Academy of Sciences* 100 (2003): 8030–35.

Goldstein, N. J., R. B. Cialdini, and V. Griskevicius, "A Room with a Viewpoint: Using Social Norms to Motivate Environmental Conservation in Hotels." *Journal of Consumer Research* 35 (2008): 472–82.

Gottlieb, G. "Normally Occurring Environmental and Behavioral Influences on Gene Activity: From Central Dogma to Probabilistic Epigenesis." *Psychological Review* 105 (1998): 792–802.

———. *Synthesizing Nature-Nurture: Prenatal Roots of Instinctive Behavior*. Mahway, NJ: Lawrence Erlbaum, 1997.

Gottman, J. M. *What Predicts Divorce? The Relationship between Marital Processes and Marital Outcomes*. Hillsdale, NJ: Erlbaum, 1994.

Gould, E. "How Widespread Is Adult Neurogenesis in Mammals?" *Nature Reviews Neuroscience* 8 (2007): 481–88.

Gould, J. "Stop Feeling Like a Fake." *Monitor on Psychology* 39, no. 7 (July 2008): 78.

Gould, J. L. "How Bees Remember Flower Shapes." *Science* 227 (1985): 1492–94.

Graham, J. M., and C. Desjardins. "Classical Conditioning: Induction of Luteinizing Hormone and Testosterone Secretion in Anticipation of Sexual Activity." *Science* 210 (1980): 1039–41.

Granier-Deferre, C., S. Bassereau, A. Ribeiro, A. Y. Jacquet, and A. J. DeCasper. "A Melodic Contour Repeatedly Experienced by Human Near-Term Fetuses Elicits a Profound Cardiac Reaction One Month after Birth." *PLoS ONE* 6 (2011): e17304.

Grant, B. R., and P. R. Grant. "What Darwin's Finches Can Teach Us about the Evolutionary Origin and Regulation of Biodiversity." *Bioscience* 53 (2003): 965–75.

Grant, B. S. "Fine Tuning the Peppered Moth Paradigm." *Evolution* 53 (1999): 980–84.

Gray, R. D. "Sparrows, Matching and the Ideal Free Distribution: Can Biological and Psychological Approaches Be Synthesized?" *Animal Behaviour* 48 (1994): 411–23.

Green, D. M., and Swets, J. A. *Signal Detection Theory and Psychophysics*. New York: Wiley, 1966.

Green, L., and J. Myerson. "A Discounting Framework for Choice with Delayed and Probabilistic Rewards." *Psychological Bulletin* 130 (2004): 769–92.

Greene, L., D. Kamps, J. Wyble, and C. Ellis. "Home-Based Consultation for Parents of Young Children with Behavioral Problems." *Child and Family Behavior Therapy* 21 (1999): 19–45.

Grescoe, T. *Bottomfeeder: How to Eat Ethically in a World of Vanishing Seafood*. New York: Bloomsbury, 2008.

Griffin, A. K. *Aristotle's Psychology of Conduct*. London: Williams and Norgate, 1931.

Grosch, J., and A. Neuringer. "Self-Control in Pigeons under the Mischel Paradigm." *Journal of the Experimental Analysis of Behavior* 35 (1981): 3–21.

Gros-Louis, J., M. H. Goldstein, A. P. King, and M. J. West. "Mothers Provide Differential Feedback to Infants' Prelinguistic Sounds." *International Journal of Behavioral Development* 30 (2006): 112–19.

Grossmann, K. E. "Continuous, Fixed-Ratio, and Fixed-Interval Reinforcement in Honey Bees." *Journal of the Experimental Analysis of Behavior* 20 (1973): 105–109.

Gubernick, D. J., and J. R. Alberts. "Maternal Licking of Young: Resource Exchange and Proximate Controls." *Physiology & Behavior* 31 (1983): 593–601.

Guo, J., J. Wang, and F. Luo. "Dissection of Placebo Analgesia in Mice: The Conditions for Activation of Opioid and Non-Opioid Systems." *Journal of Psychopharmacology* 24 (2010): 1561–67.

Hackenberg, T. D., and V. R. Joker. "Instructional versus Schedule Control of Humans' Choices in Situations of Diminishing Returns." *Journal of the Experimental Analysis of Behavior* 62 (1994): 367–83.

Hager, T. *The Demon under the Microscope: From Battlefield Hospitals to Nazi Labs, One Doctor's Heroic Search for the World's First Miracle Drug.* New York: Harmony, 2006.

Halpin, G., and G. Halpin. "Experimental Investigation of the Effects of Study and Testing on Student Learning, Retention, and Ratings of Instruction." *Journal of Educational Psychology* 74 (1982): 32–38.

Hanson, J. D., M. E. Larson, and C. T. Snowdon. "The Effects of Control over High Intensity Noise on Plasma Cortisol Levels in Rhesus Monkeys." *Behavioral Biology* 16 (1976): 333–40.

Hare, R. D. *Without Conscience: The Disturbing World of the Psychopaths among Us.* New York: Pocket, 1999.

Harlow, H. F., M. K. Harlow, and D. R. Meyer. "Learning Motivated by a Manipulation Drive." *Journal of Experimental Psychology* 40 (1950): 228–34.

Harrell, R. F., E. Woodyard, and A. I. Gates. *The Effect of Mothers' Diets on the Intelligence of the Offspring.* New York: Teachers College, 1955.

Harris, M. J., and R. Rosenthal. "Mediation of Interpersonal Expectancy Effects: 31 Meta-Analyses." *Psychological Bulletin* 97 (1985): 363–86.

Harshaw, C., I. P. Tourgeman, and R. Lickliter. "Stimulus Contingency and the Malleability of Species-Typical Auditory Preferences in Northern Bobwhite (*Colinus virginianus*) Hatchlings." *Developmental Psychobiology* 50 (2008): 460–72.

Hart, B., N. J. Reynolds, D. M. Baer, E. R. Brawley, and F. R. Harris. "Effect of Contingent and Non-Contingent Social Reinforcement on the Cooperative Play of a Preschool Child." *Journal of Applied Behavior Analysis* 1 (1968): 73–76.

Hart, B., and T. R. Risley. *Meaningful Differences in the Everyday Experience of Young American Children.* Baltimore: Paul H. Brookes, 1995.

Harwood, M., and E. Porter. *Moments of Discovery: Adventures with American Birds.* New York: Arch Cape, 1977.

Hayes, S. C., A. J. Brownstein, J. R. Haas, and D. E. Greenway. "Instructions, Multiple

Schedules, and Extinction: Distinguishing Rule-Governed from Schedule-Controlled Behavior." *Journal of the Experimental Analysis of Behavior* 46 (1986): 137–47.

Hebb, D. O. *The Organization of Behavior: A Neuropsychological Theory.* New York: Wiley, 1949.

Heckman, J. J. "The Economics, Technology, and Neuroscience of Human Capability Formation." *Proceedings of the National Academy of Sciences* 104 (2007): 13250–55.

Heckman, J. J., and D. V. Masterov. "The Productivity Argument for Investing in Young Children." *Applied Economic Perspectives and Policy* 29 (2007): 446–93.

Hefferline, R. F., and B. Keenan. "Amplitude-Induction Gradient of a Small-Scale (Covert) Operant." *Journal of the Experimental Analysis of Behavior* 6 (1963): 307–15.

Heijmans, B. T., E. W. Tobia, A. D. Stein, H. Putter, G. J. Blauw, E. S. Sussere, P. E. Slagboom, and L. H. Lumey. "Persistent Epigenetic Differences Associated with Prenatal Exposure to Famine in Humans." *Proceedings of the National Academy of Sciences* 105 (2008): 17046–49.

Heim, C., and C. B. Nemeroff. "The Role of Childhood Trauma in the Neurobiology of Mood and Anxiety Disorders: Preclinical and Clinical Studies." *Biological Psychiatry* 49 (2001): 1023–39.

Henig, J. R. *What Do We Know about the Outcomes of KIPP Schools?* East Lansing, MI: Great Lakes Center for Education Research and Practice, November 2008.

Herman, L. M., D. G. Richards, and J. P. Wolz. "Comprehension of Sentences by Bottlenosed Dolphins." *Cognition* 16 (1984): 129–219.

Herrnstein, R. J., and P. A. de Villiers. "Fish as a Natural Category for People and Pigeons." *Psychology of Learning and Motivation* 14 (1980): 59–95.

Herrnstein, R. J., D. H. Loveland, and C. Cable. "Natural Concepts in Pigeons." *Journal of Experimental Psychology: Animal Behavior Processes* 2 (1976): 285–302.

Higgins, S. T., D. D. Delaney, A. J. Budney, W. K. Bickel, J. R. Hughes, F. Foerg, and J. W. Fenwick. "A Behavioral Approach to Achieving Initial Cocaine Abstinence." *American Journal of Psychiatry* 148 (1991): 1218–24.

Higgins, S. T., E. K. Morris, and L. M. Johnson. "Social Transmission of Superstitious Behavior in Preschool Children." *Psychological Record* 39 (1989): 307–23.

Hilty, S. L. *Birds of Tropical America: A Watcher's Introduction to Behavior, Breeding, and Diversity.* Shelburne, VT: Chapters Publishing, 1994.

Hiroto, D. S. "Locus of Control and Learned Helplessness." *Journal of Experimental Psychology,* 102 (1974): 187–93.

Hoch, J., and F. J. Symons. "Matching Analysis of Socially Appropriate and Destructive Behavior in Developmental Disabilities." *Research in Developmental Disabilities* 28 (2007): 238–48.

Hochberg, L. R., M. D. Serruya, G. M. Friehs, J. A. Mukand, M. Saleh, A. H. Caplan, A. Branner, et al. "Neuronal Ensemble Control of Prosthetic Devices by a Human with Tetraplegia." *Nature* 442 (2006): 164–71.

Hockey, G. R. J., and F. Earle. "Control over the Scheduling of Simulated Office Work Reduces the Impact of Workload on Mental Fatigue and Task Performance." *Journal of Experimental Psychology: Applied* 12 (2006): 50–65.

Hoebel, B. G. "Neuroscience and Motivation: Pathways and Peptides That Define Motivational Systems." In *Stevens' Handbook of Experimental Psychology*, 2nd ed., edited by R. C. Atkinson, R. J. Herrnstein, G. Lindzey, and R. D. Luce, pp. 547–625. Oxford: John Wiley and Sons, 1988.

Hoffman, C. M., G. S. Flory, and J. R. Alberts. "Neonatal Thermotaxis Improves Reversal of a Thermally Reinforced Operant Response." *Developmental Psychobiology* 34 (1999): 87–99.

Hoffman, H. G., A. Garcia-Palacios, A. Carlin, T. A. Furness, and C. Botella-Arbona. "Interfaces That Heal: Coupling Real and Virtual Objects to Treat Spider Phobia." *International Journal of Human-Computer Interaction* 16 (2003): 283–300.

Hoffman, H. S. *Amorous Turkeys and Addicted Ducklings: The Science of Social Bonding and Imprinting*. Boston: Authors Cooperative, 1996.

Hogan, L. C., M. Bell, and R. Olson. "Preliminary Investigation of the Reinforcement Function of Signal Detections in Simulated Baggage Screening: Further Support for the Vigilance Reinforcement Hypothesis." *Journal of Organizational Behavior Management* 29 (2009): 6–18.

Hokanson, J. E., K. R. Willers, and E. Koropsak. "The Modification of Autonomic Responses during Aggressive Interchange." *Journal of Personality* 36 (1968): 386–404.

Hollard, V., and M. C. Davison. "Preference for Qualitatively Different Reinforcers." *Journal of the Experimental Analysis of Behavior* 16 (1971): 375–80.

Hollerman, J. R., and W. Schultz. "Dopamine Neurons Report an Error in the Temporal Prediction of Reward during Learning." *Nature Neuroscience* 1 (1998): 304–309.

Holz, W. C., and N. H. Azrin. "Discriminative Properties of Punishment." *Journal of the Experimental Analysis of Behavior* 4 (1961): 225–32.

———. "Interactions between the Discriminative and Aversive Properties of Punishment." *Journal of the Experimental Analysis of Behavior* 5 (1962): 229–34.

Holzhaider, J. C., G. R. Hunt, and R. D. Gray. "The Development of Pandanus Tool Manufacture in Wild New Caledonian Crows." *Behaviour* 147 (2010a): 553–86.

———. "Social Learning in New Caledonian Crows." *Learning & Behavior* 38 (2010b): 206–19.

Hopko, D. R., C. W. Lejuez, J. P. Lepage, S. D. Hopko, and D. W. McNeil. "A Brief Behavioral Activation Treatment for Depression: A Randomized Pilot Trial within an Inpatient Psychiatric Hospital." *Behavior Modification* 27 (2003): 458–69.

Horne, P. J., and C. F. Lowe. "Determinants of Human Performance on Concurrent Schedules." *Journal of the Experimental Analysis of Behavior* 59 (1993): 29–60.

House, J. S., K. R. Landis, and D. Umberson. "Social Relationships and Health." *Science* 241 (1988): 540–45.

Houston, A. "The Matching Law Applies to Wagtails' Foraging in the Wild." *Journal of the Experimental Analysis of Behavior* 45 (1986): 15–18.

Howard, K. I., S. M. Kopta, M. S. Krause, and D. E. Orlinsky. "The Dose-Effect Relationship in Psychotherapy." *American Psychologist* 41 (1986): 159–64.

Howard, M. L., and K. G. White. "Social Influence in Pigeons (*Columba livia*): The Role of Differential Reinforcement." *Journal of the Experimental Analysis of Behavior* 79 (2003): 175–91.

Hughes, B. O., and A. J. Black. "The Preference of Domestic Hens for Different Types of Battery Cage Floor." *British Poultry Science* 14 (1973): 615–19.

Hughes, R. *The Fatal Shore.* New York: Vintage Books, 1986.

Humphrey, C. R., and Krout, J. A. "Traffic and the Suburban Highway Neighbor." *Traffic Quarterly* 29 (1975): 593–613.

Isaksen, J., and P. Holth. "An Operant Approach to Teaching Joint Attention Skills to Children with Autism." *Behavioral Interventions* 24 (2009): 215–36.

Itard, J. *The Wild Boy of Aveyron.* Translated by G. Humphrey and M. Humphrey. New York: Appleton-Century-Crofts, 1962. Originally published in 1801 and in 1806.

Iversen, I. H., and J. Mogensen. "A Multipurpose Vertical Holeboard with Automated Recording of Spatial and Temporal Response Patterns for Rodents." *Journal of Neuroscience Methods* 25 (1988): 251–63.

Iwata, B. A., G. M. Pace, M. F. Dorsey, J. R. Zarcone, T. R. Vollmer, R. G. Smith, T. A. Rodgers, et al. "The Functions of Self-Injurious Behavior: An Experimental-Epidemiological Analysis." *Journal of Applied Behavior Analysis* 27 (1994): 215–40.

Iyengar, S. S., and M. R. Lepper. "When Choice Is Demotivating: Can One Desire Too Much of a Good Thing?" *Journal of Personality and Social Psychology* 79 (2000): 995–1006.

Jablonka, E., and M. J. Lamb. *Evolution in Four Dimensions: Genetic, Epigenetic, Behavioral, and Symbolic Variation in the History of Life.* Cambridge, MA: MIT Press, 2005.

Jablonka, E., and G. Raz. "Transgenerational Epigenetic Inheritance: Prevalence, Mechanisms, and Implications for the Study of Heredity and Evolution." *Quarterly Review of Biology* 84 (2009): 131–76.

Jackson, J. B. C., M. X. Kirby, W. H. Berger, K. A. Bjorndal, L. W. Botsford, B. J. Bourque, R. H. Bradbury, et al. "Historical Overfishing and the Recent Collapse of Coastal Ecosystems." *Science* 293 (2001): 629–37.

Jackson, K., and T. D. Hackenberg. "Token Reinforcement, Choice, and Self-Control in Pigeons." *Journal of the Experimental Analysis of Behavior* 66 (1996): 29–49.

Janata, P., B. Tillmann, and J. J. Bharucha. "Listening to Polyphonic Music Recruits Domain-General Attention and Working Memory Circuits." *Cognitive, Affective & Behavioral Neuroscience* 2 (2002): 121–40.

Jimura K., J. Myerson, J. Hilgard, T. S. Braver, and L. Green. "Are People Really More Patient Than Other Animals? Evidence from Human Discounting of Real Liquid Rewards." *Psychonomic Bulletin and Review* 16 (2009): 1071–75.

Johansen, E. B., P. R. Killeen, V. A. Russell, G. Tripp, J. R. Wickens, R. Tannock, J. Williams, and T. Sagvolden. "Origins of Altered Reinforcement Effects in ADHD." *Behavioral and Brain Functions* 5 (2009): 7.

Johansen-Berg, H., M. F. S. Rushworth, M. D. Bogdanovic, U. Kischka, S. Wimalaratna, and P. M. Matthews. "The Role of Ipsilateral Premotor Cortex in Hand Movement after Stroke." *Proceedings of the National Academy of Sciences* 99 (2002): 14518–23.

Johns, M., T. Schmader, and A. Martens. "Knowing Is Half the Battle: Teaching Stereotype Threat as a Means of Improving Women's Math Performance." *Psychological Science* 16 (2005): 175–79.

Johnson, D. A., and A. M. Dickinson. "Employee of the Month Programs: Do They Really Work?" *Journal of Organizational Behavior Management* 30 (2010): 308–24.

Johnson, N. A. *Darwinian Detectives: Revealing the Natural History of Genes and Genomes.* Oxford: Oxford University Press, 2007.

Jolly, A. *The Evolution of Primate Behavior.* 2nd ed. New York: Macmillan, 1985.

Jones, N. *The Plimsoll Sensation: The Great Campaign to Save Lives at Sea.* London: Little, Brown, 2006.

Jwaideh, A. R. "Responding under Chained and Tandem Fixed-Ratio Schedules." *Journal of the Experimental Analysis of Behavior* 19 (1973): 259–67.

Jwaideh, A. R., and D. E. Mulvaney. "Punishment of Observing by a Stimulus Associated with the Lower of Two Reinforcement Frequencies." *Learning and Motivation* 7 (1976): 211–22.

Kahng, S., B. A. Iwata, and A. B. Lewin. "Behavioral Treatment of Self-Injury, 1964 to 2000." *American Journal on Mental Retardation* 107 (2002): 212–21.

Karjalainen, S. "Consumer Preferences for Feedback on Household Electricity Consumption." *Energy and Buildings* 43 (2011): 458–67.

Karlsson, N., G. Loewenstein, and D. Seppi. "The Ostrich Effect: Selective Attention to Information." *Journal of Risk and Uncertainty* 38 (2009): 95–115.

Kast, B. "Decisions, Decisions." *Nature* 411 (2001): 126–28.

Kastelein, R. A., and P. R. Wiepkema. "The Significance of Training for the Behaviour of Steller Sea Lions (*Eumetopias jubatai*) in Human Care." *Aquatic Mammals* 14 (1988): 39–41.

Katz, J. S., A. A. Wright, and K. D. Bodily. "Issues in the Comparative Cognition of Abstract Concept Learning." *Comparative Cognition & Behavior Reviews* 1 (2007): 79–92.

Kavanau, J. L. "Compulsory Regime and Control of Environment in Animal Behavior: I. Wheel-Running." *Behaviour* 20 (1963): 251–81.

Kazdin, A. E. *The Kazdin Method for Parenting the Defiant Child*. Boston: Houghton Mifflin, 2008.

———. "Role of Instructions and Reinforcement in Behavior Changes in Token Reinforcement Programs." *Journal of Educational Psychology* 64 (1973): 63–71.

Kelley, A. E., and K. C. Berridge. "The Neuroscience of Natural Rewards: Relevance to Addictive Drugs." *Journal of Neuroscience* 22 (2002): 3306–11.

Kelly, J. F., and D. F. Hake. "An Extinction-Induced Increase in an Aggressive Response with Humans." *Journal of the Experimental Analysis of Behavior* 14 (1970): 153–64.

Kenison, K. Foreword to *Best American Short Stories 2003*, edited by W. Mosley, pp. v–x. Boston: Houghton Mifflin, 2003.

Kennedy, C. H., M. Caruso, and T. Thompson. "Experimental Analyses of Gene-Brain-Behavior Relations: Some Notes on Their Application." *Journal of Applied Behavior Analysis* 34 (2001): 539–49.

Kern, L., C. M. Vorndran, A. Hilt, J. E. Ringdahl, B. E. Adelman, and G. Dunlap. "Choice as an Intervention to Improve Behavior: A Review of the Literature." *Journal of Behavioral Education* 8 (1998): 151–69.

Kessel, E. L. "The Mating Activities of Balloon Flies." *Systematic Zoology* 4 (1955): 97–104.

Kessler, R. C. "The Effects of Stressful Life Events on Depression." *Annual Review of Psychology* 48 (1997): 191–214.

Kieran, J. *Footnotes on Nature*. Garden City, NY: Doubleday, 1947.

King, A. P., M. J. West, and M. H. Goldstein. "Non-Vocal Shaping of Avian Song Development: Parallels to Human Speech Development." *Ethology* 111 (2005): 101–17.

Kish, G. B., and G. W. Barnes. "Reinforcing Effects of Manipulation in Mice." *Journal of Comparative and Physiological Psychology* 54 (1961): 713–15.

Kivetz, R., O. Urminsky, and Y. Zheng. "The Goal-Gradient Hypothesis Resurrected: Purchase Acceleration, Illusionary Goal Progress, and Customer Retention." *Journal of Marketing Research* 43 (2006): 39–58.

Kjelle, M. M. *S. E. Hinton: Author of* The Outsiders. Berkeley Heights, NJ: Enslow, 2008.

Kleim, J. A., S. Barbay, N. R. Cooper, T. M. Hogg, C. N. Reidel, M. S. Remple, and R. J. Nudo. "Motor Learning-Dependent Synaptogenesis Is Localized to Functionally Reorganized Motor Cortex." *Neurobiology of Learning and Memory* 77 (2002): 63–77.

Kleim, J. A., E. Lussnig, E. R. Schwarz, T. A. Comery, and W. T. Greenough. "Synaptogenesis and FOS Expression in the Motor Cortex of the Adult Rat after Motor Skill Learning." *Journal of Neuroscience* 16 (1996): 4529–35.

Klein, J. L., R. F. P. MacDonald, G. Vaillancourt, W. H. Ahearn, and W. V. Dube. "Teaching Discrimination of Adult Gaze Direction to Children with Autism." *Research in Autism Spectrum Disorders* 3 (2009): 42–49.

Klein, K. J. K., and S. D. Hodges. "Gender Differences, Motivation, and Empathic Accuracy: When It Pays to Understand." *Personality and Social Psychology Bulletin* 27 (2001): 720–30.

Knutson, B., G. W. Fong, S. M. Bennett, C. M. Adams, and D. Hommer. "A Region of Mesial Prefrontal Cortex Tracks Monetarily Rewarding Outcomes: Characterization with Rapid Event-Related fMRI." *NeuroImage* 18 (2003): 263–72.

Koda, H., C. Oyakawa, A. Kato, and N. Masataka. "Experimental Evidence for the Volitional Control of Vocal Production in an Immature Gibbon." *Behaviour* 144 (2007): 681–92.

Kohlenberg, R., and T. Phillips. "Reinforcement and Rate of Litter Depositing." *Journal of Applied Behavior Analysis* 6 (1973): 391–96.

Kollins, S. H. "Delay Discounting Is Associated with Substance Use in College Students." *Addictive Behaviors* 28 (2003): 1167–73.

Kollins, S. H., M. C. Newland, and T. S. Critchfield. "Human Sensitivity to Reinforcement in Operant Choice: How Much Do Consequences Matter?" *Psychonomic Bulletin and Review* 4 (1997): 208–20.

Koo, M., and A. Fishbach. "Dynamics of Self-Regulation: How (Un)accomplished Goal Actions Affect Motivation." *Journal of Personality and Social Psychology* 94 (2008): 183–95.

Kortick, S. A., and R. M. O'Brien. "The World Series of Quality Control: A Case Study in the Package Delivery Industry." *Journal of Organizational Behavior Management* 16 (1996): 77–93.

Krause, T. R., K. J. Seymour, and K. C. M. Sloat. "Long-Term Evaluation of a Behavior-Based Method for Improving Safety Performance: A Meta-Analysis of 73 Interrupted Time-Series Replications." *Safety Science* 32 (1999): 1–18.

Krebs, D. "Empathy and Altruism." *Journal of Personality and Social Psychology* 32 (1975): 1134–46.

Krützen, M., J. Mann, M. R. Heithaus, R. C. Connor, L. Bejder, and W. B. Sherwin. "Cultural Transmission of Tool Use in Bottlenose Dolphins." *Proceedings of the National Academy of Sciences* 102 (2005): 8939–43.

Kuczaj, S. A. "Children's Judgments of Grammatical and Ungrammatical Irregular Past-Tense Verbs." *Child Development* 49 (1978): 319–26.

Kupfer, A. S., R. Allen, and E. F. Malagodi. "Induced Attack during Fixed-Ratio and Matched-Time Schedules of Food Presentation." *Journal of the Experimental Analysis of Behavior* 89 (2008): 31–48.

Kusayama, T., and S. Watanabe. "Reinforcing Effects of Methamphetamine in Planarians." *Neuroreport* 11 (2000): 2511–13.

LaBarbera, J. D., and R. M. Church. "Magnitude of Fear as a Function of Expected Time to an Aversive Event." *Animal Learning & Behavior* 2 (1974): 199–202.

Lamarre, J., and J. G. Holland. "The Functional Independence of Mands and Tacts." *Journal of the Experimental Analysis of Behavior* 43 (1985): 5–19.

Lamb, R. J., K. C. Kirby, A. R. Morral, G. Galbicka, and M. Y. Iguchi. "Shaping Smoking Cessation in Hard-to-Treat Smokers." *Journal of Consulting and Clinical Psychology* 78 (2010): 62–71.

Langer, E. J., and J. Rodin. "The Effects of Choice and Enhanced Personal Responsibility for the Aged: A Field Experiment in an Institutional Setting." *Journal of Personality and Social Psychology* 34 (1976): 191–98.

Lanphear, B. P., R. Hornung, J. Khoury, K. Yolton, P. Baghurst, D. C. Bellinger, R. L. Canfield, et al. "Low-Level Environmental Lead Exposure and Children's Intellectual Function: An International Pooled Analysis." *Environmental Health Perspectives* 113 (2005): 894–99.

Latané, B., and S. Nida. "Ten Years of Research on Group Size and Helping." *Psychological Bulletin* 89 (1981): 308–24.

Latham, G. I. "Interacting with At-Risk Children: The Positive Approach." *Principal* 72 (1992): 26–30.

Lattal, K. A., and S. Gleeson. "Response Acquisition with Delayed Reinforcement." *Journal of Experimental Psychology: Animal Behavior Processes* 16 (1990): 27–39.

Lattal, K. A., and B. Metzger. "Response Acquisition by Siamese Fighting Fish (*Betta splendens*) with Delayed Visual Reinforcement." *Journal of the Experimental Analysis of Behavior* 61 (1994): 35–44.

Lazic, M., S. M. Schneider, and R. Lickliter. "Enriched Rearing Facilitates Spatial Exploration in Northern Bobwhite (*Colinus virginianus*) Neonates." *Developmental Psychobiology* 49 (2007): 548–51.

Learmonth, A. E., R. Lamberth, and C. Rovee-Collier. "Generalization of Deferred Imitation during the First Year of Life." *Journal of Experimental Child Psychology* 88 (2004): 297–318.

Least Heat Moon, W. *Blue Highways: A Journey into America*. New York: Fawcett Crest, 1982.

Lee, A., S. Clancy, and A. S. Fleming. "Mother Rats Bar-Press for Pups: Effects of

Lesions of the MPOA and Limbic Sites on Maternal Behavior and Operant Responding for Pup-Reinforcement." *Behavioural Brain Research* 100 (1999): 15–31.

Leight, K. A., and H. C. Ellis. "Emotional Mood States, Strategies, and State-Dependency in Memory." *Journal of Verbal Learning and Verbal Behavior* 20 (1981): 251–66.

Lejeune, H., and J. H. Wearden. "Scalar Properties in Animal Timing: Conformity and Violations." *Quarterly Journal of Experimental Psychology* 59 (2006): 1875–1908.

Lejuez, C. W., G. H. Eifert, M. J. Zvolensky, and J. B. Richards. "Preference between Onset Predictable and Unpredictable Administrations of 20% Carbon-Dioxide-Enriched Air: Implications for Better Understanding the Etiology and Treatment of Panic Disorder." *Journal of Experimental Psychology: Applied* 6 (2000): 349–58.

Lekander, M., C. J. Fürst, S. Rotstein, H. Blomgren, and M. Fredrikson. "Anticipatory Immune Changes in Women Treated with Chemotherapy for Ovarian Cancer." *International Journal of Behavioral Medicine* 2 (1995): 1–12.

Leshan, L. L., and R. E. Worthington. "Personality as a Factor in the Pathogenesis of Cancer: A Review of the Literature." *British Journal of Medical Psychology* 29 (1956): 49–56.

Levenson, R. W., and J. M. Gottman. "Marital Interaction: Physiological Linkage and Affective Exchange." *Journal of Personality and Social Psychology* 45 (1983): 587–97.

Levey, D. J., G. A. Londoño, J. Ungvari-Martin, M. R. Hiersoux, J. E. Jankowski, J. R. Poulsen, C. M. Stracey, and S. K. Robinson. "Urban Mockingbirds Quickly Learn to Recognize Individual People." *Proceedings of the National Academy of Sciences* 106 (2009): 8959–62.

Levitt, S. D., and S. J. Dubner. *Freakonomics: A Rogue Economist Explores the Hidden Side of Everything*. New York: Morrow, 2005.

Lewontin, R. "The Units of Selection." *Annual Reviews of Ecology and Systematics* 1 (1970): 1–18.

Li, G., J. Wang, S. J. Rossiter, G. Jones, and S. Zhang. "Accelerated FoxP2 Evolution in Echolocating Bats." *PLoS ONE* 2 (2007): e900.

Li, J., J. Krauth, and J. P. Huston. "Operant Behavior of Rats under Fixed-Interval Reinforcement Schedules: A Dynamical Analysis via the Extended Return Map." *Nonlinear Dynamics, Psychology, and Life Sciences* 10 (2006): 215–40.

Lieberman, P. *Human Language and Our Reptilian Brain: The Subcortical Bases of Speech, Syntax, and Thought*. Cambridge, MA: Harvard University Press, 2002.

Lightdale, J. R., and D. A. Prentice. "Rethinking Sex Differences in Aggression:

Aggressive Behavior in the Absence of Social Roles." *Personality and Social Psychology Bulletin* 20 (1994): 34–44.

Lilly, J. S., and A. M. Miller. "Operant Conditioning of the Bottlenose Dolphin with Electrical Stimulation of the Brain." *Journal of Comparative and Physiological Psychology* 55 (1962): 73–77.

List, J. A., and D. Lucking-Reiley. "The Effects of Seed Money and Refunds on Charitable Giving: Experimental Evidence from a University Capital Campaign." *Journal of Political Economy* 110 (2002): 215–33.

Livingston, G., K. Johnston, C. Katona, J. Paton, C. G. Lyketsos, and Old Age Task Force of the World Federation of Biological Psychiatry. "Systematic Review of Psychological Approaches to the Management of Neuropsychiatric Symptoms of Dementia." *American Journal of Psychiatry* 162 (2005): 1996–2021.

Lockhart, R. A. "Temporal Conditioning of GSR." *Journal of Experimental Psychology* 71 (1966): 438–46.

Loewenstein, G. "Anticipation and the Valuation of Delayed Consumption." *Economic Journal* 97 (1987): 666–84.

Logue, A. W. "Taste Aversion and the Generality of the Laws of Learning." *Psychological Bulletin* 86 (1979): 276–96.

Lohr, J. M., B. O. Olatunji, and C. N. Sawchuk. "A Functional Analysis of Danger and Safety Signals in Anxiety Disorders." *Clinical Psychology Review* 27 (2007): 114–26.

Lorenzetti, F. D., R. Mozzachiodi, D. A. Baxter, and J. H. Byrne. "Classical and Operant Conditioning Differentially Modify the Intrinsic Properties of an Identified Neuron." *Nature Neuroscience* 9 (2005): 17–19.

Losada, M., and E. Heaphy. "The Role of Positivity and Connectivity in the Performance of Business Teams: A Nonlinear Dynamics Model." *American Behavioral Scientist* 47 (2004): 740–65.

Louie, K., and P. W. Glimcher. "Separating Value from Choice: Delay Discounting Activity in the Lateral Intraparietal Area." *Journal of Neuroscience* 30 (2010): 5498–507.

Louie, T. A., R. L. Kulik, and R. Jacobson. "When Bad Things Happen to the Endorsers of Good Products." *Marketing Letters* 12 (2001): 13–23.

Lovaas, O. I. "Behavioral Treatment and Normal Educational and Intellectual Functioning in Young Autistic Children." *Journal of Consulting and Clinical Psychology* 55 (1987): 3–9.

Lowe, C. F., A. Beasty, and R. P. Bentall. "The Role of Verbal Behavior in Human Learning: Infant Performance on Fixed-Interval Schedules." *Journal of the Experimental Analysis of Behavior* 39 (1983): 157–64.

Lowe, C. F., P. J. Horne, K. Tapper, M. Bowdery and C. Egerton. "Effects of a Peer Modelling and Rewards-Based Intervention to Increase Fruit and Vegetable Consumption in Children." *European Journal of Clinical Nutrition* 58 (2004): 510–22.

Ludwig, T. D., and E. S. Geller. "Intervening to Improve the Safety of Delivery Drivers: A Systematic Behavioral Approach." *Journal of Organizational Behavior Management* 19 (2000): 1–124.

Lyko, F., S. Foret, R. Kucharski, S. Wolf, C. Falckenhayn, and R. Maleszka. "The Honey Bee Epigenomes: Differential Methylation of Brain DNA in Queens and Workers." *PLoS Biology* 8 (2010): e1000506.

Lyn, H., P. M. Greenfield, S. Savage-Rumbaugh, K. Gillespie-Lynch, and W. D. Hopkins. "Nonhuman Primates Do Declare! A Comparison of Declarative Symbol and Gesture Use in Two Children, Two Bonobos, and a Chimpanzee." *Language & Communication* 31 (2011): 63–74.

Lyon, C. "Doing a Rowling." In *Jeff Herman's Guide to Book Publishers, Editors, and Literary Agents 2008*, edited by J. Herman, pp. 861–64. Stockbridge, MA: Three Dog Press, 2008.

Macario, J. F. "Young Children's Use of Color in Classification: Foods and Canonically Colored Objects." *Cognitive Development* 6 (1991): 17–46.

Maccoby, E. E. *The Two Sexes: Growing Up Apart, Coming Together*. Cambridge, MA: Harvard University Press, 2003.

Mace, F. C., B. McCurdy, and E. A. Quigley. "A Collateral Effect of Reward Predicted by Matching Theory." *Journal of Applied Behavior Analysis* 23 (1990): 197–205.

Machado, A., R. Keen, and E. Macaux, "Making Analogies Work: A Selectionist Model of Choice Behavior." In *Reflections on Adaptive Behavior: Essays in Honor of J. E. R. Staddon*, edited by N. K. Innis, pp. 23–50. Cambridge, MA: MIT Press, 2009.

MacQueen G., J. Marshall, M. Perdue, S. Siegel, and J. Bienenstock. "Pavlovian Conditioning of Rat Mucosal Mast Cells to Secrete Rat Mast Cell Protease II." *Science* 243 (1989): 83–85.

Madden, G. J., and M. Perone. "Human Sensitivity to Concurrent Schedules of Reinforcement: Effects of Observing Schedule-Correlated Stimuli." *Journal of the Experimental Analysis of Behavior* 71 (1999): 303–18.

Madsen, C. H., W. C. Becker, and D. R. Thomas. "Rules, Praise, and Ignoring: Elements of Elementary Classroom Control." *Journal of Applied Behavior Analysis* 1 (1968): 139–50.

Madsen, C. H., C. K. Madsen, R. A. Saudargas, W. R. Hammond, J. B. Smith, and D. E. Edgar. "Classroom Raid (Rules, Approval, Ignore, Disapproval): A Cooperative Approach for Professionals and Volunteers." *Journal of School Psychology* 8 (1970): 180–85.

Magoon, M. A., and T. S. Critchfield. "Concurrent Schedules of Positive and Negative Reinforcement: Differential-Impact and Differential-Outcomes Hypotheses." *Journal of the Experimental Analysis of Behavior* 90 (2008): 1–22.

Maher, E. "Use of Food Items for Environmental Enrichment in a Large Aviary at the San Diego Zoo." *Proceedings of the Seventh International Conference on Environmental Enrichment*, pp. 329–31. New York: Shape of Enrichment, 2005.

Maier, S. F., C. Anderson, and D. A. Lieberman. "Influence of Control of Shock on Subsequent Shock-Elicited Aggression." *Journal of Comparative and Physiological Psychology* 81 (1972): 94–100.

Maloney, R. F., and I. G. McLean. "Historical and Experimental Learned Predator Recognition in Free-Living New Zealand Robins." *Animal Behaviour* 50 (1995): 1193–201.

Manabe, K., T. Kawashima, and J. E. Staddon. "Differential Vocalization in Budgerigars: Towards an Experimental Analysis of Naming." *Journal of the Experimental Analysis of Behavior* 63 (1995): 111–26.

Maple, T. L. "Toward a Science of Welfare for Animals in the Zoo." *Journal of Applied Animal Welfare Science* 10 (2007): 63–70.

Maratsos, M. "Some Current Issues in the Study of the Acquisition of Grammar." In *Handbook of Child Psychology*, vol. 3, edited by P. H. Mussen, pp. 707–86. New York: Wiley, 1983.

Mark, E. "Reducing Stress in Northern Bald Ibis through Training." *ABMA Wellspring* 8 (2007):10–18.

Markarian, M. "Positive Parenting: How to Encourage Good Behavior." *Healthy Children Magazine* (Winter 2008): 22–23.

Markham, J. A., and W. T. Greenough. "Experience-Driven Brain Plasticity: Beyond the Synapse." *Neuron Glia Biology* 1 (2004): 351–63.

Markowitz, H. "Engineering Environments for Behavioral Opportunities in the Zoo." *Behavior Analyst* 1 (1978): 34–47.

Markowitz, H., and S. Line. "Primate Research Models and Environmental Enrichment." In *Housing, Care and Psychological Well-Being of Captive and Laboratory Primates*, edited by E. F. Segal, pp. 203–12. Park Ridge, NJ: Noyes, 1989.

Marley, E., and W. H. Morse. "Operant Conditioning in the Newly Hatched Chicken." *Journal of the Experimental Analysis of Behavior* 9 (1966): 95–103.

Marmot, M. G., G. Davey Smith, S. Stansfield, C. Patel, F. North, J. Head, I. White, E. Brunner, and A. Feeney. "Inequalities in Health 20 Years On: The Whitehall II Study of British Civil Servants." *Lancet* 337 (1991): 1387–93.

Marmot, M. G., G. Rose, M. Shipley, and P. J. Hamilton. "Employment Grade and Coronary Heart Disease in British Civil Servants." *Journal of Epidemiology and Community Health* 32 (1978): 244–49.

Marr, M. J. "Behavior Dynamics: One Perspective." *Journal of the Experimental Analysis of Behavior* 57 (1992): 249–66.

Marzluff, J. M., J. Walls, H. N. Cornell, J. C. Withey, and D. P. Craig. "Lasting Recognition of Threatening People by Wild American Crows." *Animal Behaviour* 79 (2010): 699–707.

Mataga, N., S. Fujishima, B. G. Condie, and T. K. Hensch. "Experience-Dependent Plasticity of Mouse Visual Cortex in the Absence of the Neuronal Activity-Dependent Marker egr1/zif268." *Journal of Neuroscience* 21 (2001): 9724–32.

Mattaini, M. *Peace Power for Adolescents: Strategies for a Culture of Nonviolence.* Washington, DC: NASW Press, 2001.

Matthews, L. R., and W. Temple. "Concurrent Schedule Assessment of Food Preference in Cows." *Journal of the Experimental Analysis of Behavior* 32 (1979): 245–54.

Maurice, C. *Let Me Hear Your Voice: A Family's Triumph over Autism.* New York: Knopf, 1993.

Mayer, G. R., L. K. Mitchell, T. Clementi, E. Clement-Robertson, R. Myatt, and D. Thomas Bullara. "A Dropout Prevention Program for At-Risk High School Students: Emphasizing Consulting to Promote Positive Classroom Climates." *Education & Treatment of Children* 16 (1993): 135–46.

Mazur, J. E. *Learning and Behavior.* 6th ed. Englewood Cliffs, NJ: Prentice-Hall, 2005.

———. "Procrastination by Pigeons: Preference for Larger, More Delayed Work Requirements." *Journal of the Experimental Analysis of Behavior* 65 (1996): 159–71.

Mazur, J. E., and D. R. Biondi. "Delay-Amount Tradeoffs in Choices by Pigeons and Rats: Hyperbolic versus Exponential Discounting." *Journal of the Experimental Analysis of Behavior* 91 (2009): 197–211.

Mazzucchelli, T., R. Kane, and C. Rees. "Behavioral Activation Treatments for Depression in Adults: A Meta-Analysis and Review." *Clinical Psychology: Science and Practice* 16 (2009): 383–411.

McAdie, T. M., T. M. Foster, and W. Temple. "Concurrent Schedules: Quantifying the Aversiveness of Noise." *Journal of the Experimental Analysis of Behavior* 65 (1996): 37–55.

McAdie, T. M., L. J. Keeling, H. J. Blokhuis, and R. B. Jones. "Reduction in Feather Pecking and Improvement of Feather Condition with the Presentation of a String Device to Chickens." *Applied Animal Behaviour Science* 93 (2005): 67–80.

McClellan, J. M. "Left Seat: The Psychology of Safety." *Flying* 137, June 6, 2010, pp. 8–10.

McDowell, J. J., R. Bass, and R. Kessel. "A New Understanding of the Foundation of Linear System Analysis and an Extension to Nonlinear Cases." *Psychological Review* 100 (1993): 407–19.

McDowell, J. J., M. L. Caron, S. Kulubekova, and J. P. Berg. "A Computational Theory of Selection by Consequences Applied to Concurrent Schedules." *Journal of the Experimental Analysis of Behavior* 90 (2008): 387–403.

McKenzie, B., and R. H. Day. "Orientation Discrimination in Infants: A Comparison of Visual Fixation and Operant Training Methods." *Journal of Experimental Child Psychology* 11 (1971): 366–75.

McMillan, D. E., H. Li, and C. C. Hardwick. "Schedule Control of Quantal and Graded Dose-Effect Curves in a Drug-Drug-Saline Discrimination." *Pharmacology Biochemistry and Behavior* 68 (2001): 395–402.

Medina, M., K. Jones, C. Vitale, and P. Reiser. "The Bronx Zoo's Tiger Mountain: An Exhibit as Enrichment." *Proceedings of the Seventh International Conference on Environmental Enrichment*, pp. 55–59. New York: Shape of Enrichment, 2005.

Mehrabian, A. *Silent Messages*. Oxford: Wadsworth, 1971.

Merabet, L. B., R. Hamilton, G. Schlaug, J. D. Swisher, E. T. Kiriakopoulos, N. B. Pitskel, T. Kauffman, and A. Pascual-Leone. "Rapid and Reversible Recruitment of Early Visual Cortex for Touch." *PLoS ONE* 3 (2008): e3046.

Merzenich, M. M., and R. C. deCharms, "Neural Representations, Experience, and Change." In *The Mind-Brain Continuum: Sensory Processes*, edited by R. R. Llinás and P. S. Churchland, pp. 61–82. Cambridge, MA: MIT Press, 1996.

Meyer, M. E. "Light Onset or Offset Contingencies within a Simple or Complex Environment." *Journal of Comparative and Physiological Psychology* 66 (1968): 542–44.

Michael, J. "Comments by the Discussant." *Psychological Record* 37 (1987): 37–42.

Midlarsky, E., and J. H. Bryan. "Training Charity in Children." *Journal of Personality and Social Psychology* 5 (1967): 408–15.

Milgram, S. *Obedience to Authority: An Experimental View*. New York: Harper and Row, 1974.

Miller, N. E. "Learning Resistance to Pain and Fear: Effects of Overlearning, Exposure, and Rewarded Exposure in Context." *Journal of Experimental Psychology* 60 (1960): 137–45.

Miller, R. L., P. Brickman, and D. Bolen. "Attribution versus Persuasion as a Means for Modifying Behavior." *Journal of Personality and Social Psychology* 31 (1975): 430–41.

Mineka, S., and R. Zinbarg. "A Contemporary Learning Theory Perspective on the Etiology of Anxiety Disorders: It's Not What You Thought It Was." *American Psychologist* 61 (2006): 10–26.

Minta, S. C., K. A. Minta, and D. F. L. Hunting. "Associations between Badgers (*Taxidea taxus*) and Coyotes (*Canis latrans*)." *Journal of Mammalogy* 73 (1992): 814–20.

Mischel, W., and E. B. Ebbesen. "Attention in Delay of Gratification." *Journal of Personality and Social Psychology* 16 (1970): 329–37.

Moerk, E. L. "Three-Term Contingency Patterns in Mother-Child Verbal Interactions during First-Language Acquisition." *Journal of the Experimental Analysis of Behavior* 54 (1990): 293–305.

Moffitt, T. E., L. Arseneault, D. Belsky, N. Dickson, R. J. Hancox, H. Harrington, R. Houts, et al. "A Gradient of Childhood Self-Control Predicts Health, Wealth, and Public Safety." *Proceedings of the National Academy of Sciences* 108 (2011): 2693–98.

Moir, J. *Return of the Condor*. Guilford, CT: Lyons Press, 2006.

Molina, B. S. G., S. P. Hinshaw, J. M. Swanson, L. E. Arnold, B. Vitiello, P. S. Jensen, J. N. Epstein, et al. "The MTA at 8 Years: Prospective Follow-Up of Children Treated for Combined Type ADHD in a Multisite Study." *Journal of the American Academy of Child and Adolescent Psychiatry* 48 (2009): 484–500.

Monaco, K. A., L. Olsson, and J. Hentges. "Hours of Sleep and Fatigue in Motor Carriage." *Contemporary Economic Policy* 23 (2005): 615–24.

Montaigne, Michel de. "On Vanity." In *The Essays of Montaigne*, vol. 17, edited by W. C. Hazilitt. Translated by C. Cotton. Original work published in 1580; Cotton translation published in 1877. Project Gutenberg EBook released on September 17, 2006. http://www.gutenberg.org/files/3597/3597.txt (accessed December 27, 2011).

Monterosso, J., and G. Ainslie. "The Picoeconomic Approach to Addictions: Analyzing the Conflict of Successive Motivational States." *Addiction Research and Theory* 17 (2009): 115–34.

Moore, C. L. "The Role of Maternal Stimulation in the Development of Sexual Behavior and Its Neural Basis." *Annals of the New York Academy of Sciences* 662 (1992): 160–77.

Moore, D. S. *The Dependent Gene: The Fallacy of "Nature vs. Nurture."* New York: Freeman, 2001.

Moore, T. "Bloopers." *Washington Post*, January 9, 1977.

Moradi, F., G. T. Buracas, and R. B. Buxton. "Attention Strongly Increases Oxygen Metabolic Response to Stimulus in Primary Visual Cortex." *NeuroImage* 59 (2012): 601–607.

Morange, M. *The Misunderstood Gene*. Translated by M. Cobb. Cambridge, MA: Harvard University Press, 2001.

Morgan, H. D., H. G. E. Sutherland, D. I. K. Martin, and E. Whitelaw. "Epigenetic Inheritance at the Agouti Locus in the Mouse." *Nature Genetics* 23 (1999): 314–18.

Morgan, M. "Reward-Induced Decrements and Increments in Intrinsic Motivation." *Review of Educational Research* 54 (1984): 15–30.

Morin, P. A., F. I. Archer, A. D. Foote, J. Vilstrup, E. E. Allen, P. Wade, J. Durban, et al. "Complete Mitochondrial Genome Phylogeographic Analysis of Killer Whales (*Orcinus orca*) Indicates Multiple Species." *Genome Research* 20 (2010): 908–16.

Moskowitz, B. A. "The Acquisition of Language." *Scientific American* 239 (1978): 92–108.

Mueller, C. M., and C. S. Dweck. "Praise for Intelligence Can Undermine Children's Motivation and Performance." *Journal of Personality and Social Psychology* 75 (1998): 33–52.

Muir, J. *The Wilderness World of John Muir.* Edited by E. W. Teale. Boston: Houghton Mifflin, 1954. Original work published in 1909.

Mundy, P., M. Sigman, J. Ungerer, and T. Sherman. "Defining the Social Deficits of Autism: The Contribution of Non-Verbal Communication Measures." *Journal of Child Psychology and Psychiatry* 27 (1986): 657–69.

Muraco, H. S., and Stamper M. A. "Training Spotted Eagle Rays (*Aetobatus narinari (Euphrasen)*) to Decrease Aggressive Behaviors towards Divers." *Journal of Aquariculture and Aquatic Science* 8 (2003): 88–98.

Murphy, M. S., and R. G. Cook. "Absolute and Relational Control of a Sequential Auditory Discrimination by Pigeons (*Columba livia*)." *Behavioural Processes* 77 (2008): 210–22.

Murphy, R. C. *Logbook for Grace: Whaling Brig Daisy, 1912–1913.* New York: Macmillan, 1947.

Myers, E. *When Parents Die.* Rev. ed. New York: Penguin, 1997.

Myers, S. M., and C. P. Johnson. "Management of Children with Autism Spectrum Disorders." *Pediatrics* 120 (2007): 1162–82.

Nakamichi, M., E. Kato, Y. Kojima, and N. Itoigawa. "Carrying and Washing of Grass Roots by Free-Ranging Japanese Macaques at Katsuyama." *Folia Primatologica* 69 (1998): 35–40.

Nargeot, R., and J. Simmers. "Neural Mechanisms of Operant Conditioning and Learning-Induced Behavioral Plasticity in *Aplysia*." *Cellular and Molecular Life Sciences* 68 (2011): 803–16.

Nation, J. R., and P. Massad. "Persistence Training: A Partial Reinforcement Procedure for Reversing Learned Helplessness and Depression." *Journal of Experimental Psychology: General* 107 (1978): 436–45.

Neef, N. A., F. C. Mace, M. C. Shea, and D. Shade. "Effects of Reinforcer Rate and Reinforcer Quality on Time Allocation: Extensions of Matching Theory to Educational Settings." *Journal of Applied Behavior Analysis* 25 (1992): 691–99.

Nelson, M. E. *Strong Women Stay Young.* New York: Bantam, 1997.

Nestle, M. *What to Eat.* New York: North Point Press, 2006.

Neumann, D. L. "The Effects of Physical Context Changes and Multiple Extinction Contexts on Two Forms of Renewal in a Conditioned Suppression Task with Humans." *Learning and Motivation* 37 (2006): 149–75.

Neuringer, A. "Can People Behave 'Randomly'?: The Role of Feedback." *Journal of Experimental Psychology: General* 115 (1986): 62–75.

———. "Reinforced Variability in Animals and People: Implications for Adaptive Action." *American Psychologist* 59 (2004): 891–906.

Neuringer, A., C. Deiss, and S. Imig. "Comparing Choices and Variations in People and Rats: Two Teaching Experiments." *Behavior Research Methods, Instruments & Computers* 32 (2000): 407–16.

Nishida, T. "Local Traditions and Cultural Transmission." In *Primate Societies*, edited by B. B. Smuts, D. L. Cheney, R. M. Seyfarth, R. W. Wrangham, and T. T. Struhsaker, pp. 462–74. Chicago: University of Chicago Press, 1987.

Niznikiewicz, M. A., and M. R. Delgado. "Two Sides of the Same Coin: Learning via Positive and Negative Reinforcers in the Human Striatum." *Developmental Cognitive Neuroscience* 1 (2011): 494–505.

Nuttin, J. M. "Affective Consequences of Mere Ownership: The Name Letter Effect in Twelve European Languages." *European Journal of Social Psychology* 17 (1987): 381–402.

O'Kane, G., E. A. Kensinger, and S. Corkin. "Evidence for Semantic Learning in Profound Amnesia: An Investigation with Patient H. M." *Hippocampus* 14 (2004): 417–25.

Okouchi, H. "Response Acquisition by Humans with Delayed Reinforcement." *Journal of the Experimental Analysis of Behavior* 91 (2009): 377–90.

Olds, J., and P. M. Milner. "Positive Reinforcement Produced by Electrical Stimulation of Septal Area and Other Regions of Rat Brain." *Journal of Comparative and Physiological Psychology* 47 (1954): 419–27.

O'Leary, K. D., K. F. Kaufman, R. E. Kass, and R. S. Drabman. "The Effects of Loud and Soft Reprimands on Behavior of Disruptive Students." *Exceptional Children* 37 (1970): 145–55.

Olson, M. A., and R. H. Fazio. "Implicit Attitude Formation through Classical Conditioning." *Psychological Science 12* (2001): 413–17.

Olsson, A. S., and K. Dahlborn. "Improving Housing Conditions for Laboratory Mice: A Review of 'Environmental Enrichment.'" *Laboratory Animals* 36 (2002): 243–70.

Ono, K. "Superstitious Behavior in Humans." *Journal of the Experimental Analysis of Behavior* 47 (1987): 261–71.

Orlean, S. *The Orchid Thief*. New York: Random House, 1998.

Orwell, G. "How the Poor Die." In *Shooting an Elephant and Other Essays*. New York: Harcourt, Brace, 1950.

Ostaszewski, P., L. Green, and J. Myerson. "Effects of Inflation on the Subjective Value of Delayed and Probabilistic Rewards." *Psychonomic Bulletin and Review* 5 (1998): 324–33.

Ostlund, S. B., and B. W. Balleine. "Orbitofrontal Cortex Mediates Outcome Encoding in Pavlovian but Not Instrumental Conditioning." *Journal of Neuroscience* 27 (2007): 4819–25.

Ostrom, E. "A Diagnostic Approach for Going beyond Panaceas." *Proceedings of the National Academy of Sciences* 104 (2007): 15181–87.

———. "A General Framework for Analyzing Sustainability of Social-Ecological Systems." *Science* 325 (2009): 419–22.

Padilla, A. M., C. Padilla, T. Ketterer, and D. Giacolone. "Inescapable Shocks and Subsequent Avoidance Conditioning in Goldfish (*Carrasius auratus*)." *Psychonomic Science* 20 (1970): 295–96.

Page, S., and A. Neuringer. "Variability Is an Operant." *Journal of Experimental Psychology: Animal Behavior Processes* 11 (1985): 429–52.

Palameta, B., and L. Lefebvre. "The Social Transmission of a Food-Finding Technique in Pigeons: What Is Learned?" *Animal Behaviour* 33 (1985): 892–96.

Parsons, H. M. "What Caused the Hawthorne Effect?" *Administration and Society* 10 (1978): 259–83.

Passarelli, F., A. Merante, F. E. Pontieri, V. Margotta, G. Venturini, and G. Palladini. "Opioid-Dopamine Interaction in Planaria: A Behavioral Study." *Comparative Biochemistry and Physiology* 124 (1999): 51–55.

Patel, A. D., J. R. Iversen, M. R. Bregman, and I. Schulz, "Experimental Evidence for Synchronization to a Musical Beat in a Nonhuman Animal." *Current Biology* 19 (2009): 827–30.

Pedalino, E., and V. U. Gamboa. "Behavior Modification and Absenteeism: Intervention in One Industrial Setting." *Journal of Applied Psychology* 59 (1974): 694–98.

Pedersen, L. J., L. Holm, M. B. Jensen, and E. Jorgensen. "The Strength of Pigs' Preferences for Different Rooting Materials Measured Using Concurrent Schedules of Reinforcement." *Applied Animal Behaviour Science* 94 (2005): 31–48.

Pelham, W. E. "The NIMH Multimodal Treatment Study for Attention-Deficit Hyperactivity Disorder: Just Say Yes to Drugs Alone?" *Canadian Journal of Psychiatry* 44 (1999): 981–90.

Pennebaker, J. W. "Writing about Emotional Experiences as a Therapeutic Process." *Psychological Science* 8 (1997): 162–66.

Perone, M. "Negative Effects of Positive Reinforcement." *Behavior Analyst* 26 (2003): 1–14.

Perone, M., and K. Courtney. "Fixed-Ratio Pausing: Joint Effects of Past Reinforcer Magnitude and Stimuli Correlated with Upcoming Magnitude." *Journal of the Experimental Analysis of Behavior* 57 (1992): 33–46.

Perrin, N. *Giving Up the Gun: Japan's Reversion to the Sword, 1543–1879.* Boston: D. R. Godine, 1979.

Perry, J. L., E. B. Larson, J. P. German, G. J. Madden, and M. E. Carroll. "Impulsivity (Delay Discounting) as a Predictor of Acquisition of IV Cocaine Self-Administration in Female Rats." *Psychopharmacology* 178 (2005): 193–201.

Pessotti, I. "Discrimination with Light Stimuli and a Lever-Pressing Response in *Melipona rufiventris*." *Journal of Apicultural Research* 11 (1972): 89–93.

Peterson, J. *Splendid Soups*. New York: Wiley, 2001.

Peterson, N. "Control of Behavior by Presentation of an Imprinted Stimulus." *Science* 132 (1960): 1395–96.

Petry, N. M., S. M. Alessi, J. Marx, M. Austin, and M. Tardif. "Vouchers versus Prizes: Contingency Management Treatment of Substance Abusers in Community Settings." *Journal of Consulting and Clinical Psychology* 73 (2005): 1005–14.

Petursdottir, A. I., J. E. Carr, and J. Michael. "Emergence of Mands and Tacts of Novel Objects among Preschool Children." *Analysis of Verbal Behavior* 21 (2005): 59–74.

Pfungst, O. *Clever Hans (The Horse of Mr. von Osten): A Contribution to Experimental Animal and Human Psychology*. Translated by C. L. Rahn. New York: Henry Holt, 1911. Originally published in 1907.

Phelan, T. W. *1-2-3 Magic: Effective Discipline for Children 2–12*. 3rd ed. Glen Ellyn, IL: ParentMagic, 2003.

Piacentini, J., D. W. Woods, L. Scahill, S. Wilhelm, A. L. Peterson, S. Chang, G. S. Ginsburg, et al. "Behavior Therapy for Children with Tourette Disorder: A Randomized Controlled Trial." *Journal of the American Medical Association* 303 (2010): 1929–37.

Pilegaard, H., B. Saltin, and P. D. Neufer. "Exercise Induces Transient Transcriptional Activation of the PGC-1alpha Gene in Human Skeletal Muscle." *Journal of Physiology* 546 (2003): 851–58.

Pilley, J. E., and A. K. Reid. "Border Collie Comprehends Object Names as Verbal Referents." *Behavioural Processes* 86 (2011): 184–95.

Pinaud, R. "Experience-Dependent Immediate Early Gene Expression in the Adult Central Nervous System: Evidence from Enriched-Environment Studies." *International Journal of Neuroscience* 114 (2004): 321–33.

Platt, M. L., and P. W. Glimcher. "Neural Correlates of Decision Variables in the Parietal Cortex." *Nature* 400 (1999): 233–38.

Pogrebin, L. C. *Getting Over Getting Older*. Boston: Little, Brown, 1996.

Poling, A., B. Weetjens, C. Cox, N. W. Beyene, H. Bach, and A. Sully. "Using Trained Pouched Rats to Detect Land Mines: Another Victory for Operant Conditioning." *Journal of Applied Behavior Analysis* 44 (2011): 351–55.

Polivy, J., and C. P. Herman. "If at First You Don't Succeed: False Hopes of Self-Change." *American Psychologist* 57 (2002): 677–89.

Porter, D., and A. Neuringer. "Music Discriminations by Pigeons." *Journal of Experimental Psychology: Animal Behavior Processes* 10 (1984): 138–48.

Poulson, C. L. "Differential Reinforcement of Other Than Vocalization as a Control Procedure in the Conditioning of Infant Vocalization Rate." *Journal of Experimental Child Psychology* 36 (1983): 471–89.

Poulson, C. L., E. Kymissis, K. F. Reeve, M. Andreatos, and L. Reeve. "Generalized Vocal Imitation in Infants." *Journal of Experimental Child Psychology* 51 (1991): 267–79.

Poulson, C. L., N. Kyparissos, M. Andreatos, E. Kymissis, and M. Parnes. "Generalized Imitation within Three Response Classes in Typically Developing Infants." *Journal of Experimental Child Psychology* 81 (2002): 341–57.

Pouthas, V. "Adaptation to Duration in the 2 to 5 Year Old Child/Adaptation à la Durée chez L'enfant de 2 à 5 Ans." *L'année Psychologique* 81 (1981): 33–50.

Pratkanis, A., and E. Aronson. *Age of Propaganda: The Everyday Use and Abuse of Persuasion.* New York: Freeman, 1991.

Prelec, D., and D. Simester. "Always Leave Home without It: A Further Investigation of the Credit Card Effect on Willingness to Pay." *Marketing Letters* 12 (2001): 5–12.

Prendergast, M., D. Podus, J. Finney, L. Greenwell, and J. Roll. "Contingency Management for Treatment of Substance Use Disorders: A Meta-Analysis." *Addiction* 101 (2006): 1546–60.

Provost, G. *Make Your Words Work.* Cincinnati, OH: Writer's Digest, 1990.

Pruetz, J. D., and P. Bertolani. "Savanna Chimpanzees, *Pan troglodytes verus*, Hunt with Tools." *Current Biology* 17 (2007): 412–17.

Pryor, E. B. *Clara Barton: Professional Angel.* Philadelphia: University of Pennsylvania Press, 1987.

Pryor, K. "Behavior and Learning in Porpoises and Whales." *Naturwissenschaften* 60 (1973): 412–20.

———. *Don't Shoot the Dog! The New Art of Teaching and Training* . Rev. ed. New York: Bantam, 1999.

———. *Reaching the Animal Mind: Clicker Training and What It Teaches Us about All Animals.* New York: Scribner, 2009.

———. "The Rhino Likes Violets." *Psychology Today*, April 1981, pp. 92–98.

Pryor, K., R. Haag, and J. O'Reilly. "The Creative Porpoise: Training for Novel Behavior." *Journal of the Experimental Analysis of Behavior* 12 (1969): 653–61.

Pryor, K., J. Lindbergh, S. Lindbergh, and R. A Milano. "A Dolphin-Human Fishing Cooperative in Brazil." *Marine Mammal Science* 6 (1990): 77–82.

Putnam, R. D. *Bowling Alone: The Collapse and Revival of American Community.* New York: Simon and Schuster, 2000.

Putnam, R. F., J. K. Luiselli, K. Sennett, and J. Malonson. "Cost-Efficacy Analysis of Out-of-District Special Education Placements: An Evaluative Measure of Behavior Support Intervention in Public Schools." *Journal of Special Education Leadership* 15 (2002): 17–24.

Quay, H. "The Effect of Verbal Reinforcement on the Recall of Early Memories." *Journal of Abnormal and Social Psychology* 59 (1959): 254–57.

Rachlin, H., and L. Green. "Commitment, Choice and Self-Control." *Journal of the Experimental Analysis of Behavior* 17 (1972): 15–22.

Raine, N. E., and L. Chittka. "The Correlation of Learning Speed and Natural Foraging Success in Bumble-bees." *Proceedings of the Royal Society B* 275 (2008): 803–808.

Rajala, A. K., and D. A. Hantula. "Towards a Behavioral Ecology of Consumption: Delay-Reduction Effects on Foraging in a Simulated Internet Mall." *Managerial and Decision Economics* 21 (2000): 145–58.

Rapanelli, M., S. E. Lew, L. R. Frick, and B. S. Zanutto. "Plasticity in the Rat Prefrontal Cortex: Linking Gene Expression and an Operant Learning with a Computational Theory." *PLoS ONE* 5 (2010): e8656.

Rasey, H. W., and I. H. Iversen. "An Experimental Acquisition of Maladaptive Behavior by Shaping." *Journal of Behavior Therapy and Experimental Psychiatry* 24 (1993): 37–43.

Rashotte, M. E., D. F. Foster, and T. Austin. "Two-Pan and Operant Lever-Press Tests of Dogs' Preference for Various Foods." *Neuroscience & Biobehavioral Reviews* 8 (1984): 231–37.

Rath, T., and D. Clifton. *How Full Is Your Bucket?* Omaha, NE: Gallup Press, 2004.

Razran, G. "A Direct Laboratory Comparison of Pavlovian Conditioning and Traditional Associative Learning." *Journal of Abnormal and Social Psychology* 51 (1955): 649–52.

Recanzone, G. H., M. M. Merzenich, W. M. Jenkins, K. A. Grajski, and H. R. Dinse. "Topographic Reorganization of the Hand Representation in Cortical Area 3b of Owl Monkeys Trained in a Frequency-Discrimination Task." *Journal of Neurophysiology* 67 (1992): 1031–56.

Recanzone, G. H., C. E. Schreiner, and M. M. Merzenich. "Plasticity in the Frequency Representation of Primary Auditory Cortex following Discrimination Training in Adult Owl Monkeys." *Journal of Neuroscience* 13 (1993): 87–103.

Reed, D. D., T. S. Critchfield, and B. K. Martens. "The Generalized Matching Law in Elite Sport Competition: Football Play Calling as Operant Choice." *Journal of Applied Behavior Analysis* 39 (2006): 281–97.

Reese, E. P. "The Role of Husbandry in Promoting the Welfare of Laboratory Animals." In *Animals in Biomedical Research*, edited by C. F. M. Hendriksen and H. B. W. M. Koeter, pp. 155–92. Amsterdam: Elsevier, 1991.

Reinhardt, V. "Working with Rather Than against Macaques during Blood Collection." *Journal of Applied Animal Welfare Science* 6 (2003): 189–97.

Reschly, D. J. "School Psychology Paradigm Shift and Beyond." In *Best Practices in School Psychology*, vol. 1, 5th ed., edited by A. Thomas and J. Grimes, pp. 3–15. Bethesda, MD: National Association of School Psychologists, 2008.

Reynolds, B. "A Review of Delay-Discounting Research with Humans: Relations to Drug Use and Gambling." *Behavioural Pharmacology* 17 (2006): 651–67.

Reynolds, G. S. "Attention in the Pigeon." *Journal of the Experimental Analysis of Behavior* 4 (1961): 203–208.

Richardson, D. S., and G. S. Hammock. "Social Context of Human Aggression: Are We Paying Too Much Attention to Gender?" *Aggression and Violent Behavior* 12 (2007): 417–26.

Richmond, V. P., and J. C. McCroskey. *Nonverbal Behavior in Interpersonal Relations.* Boston: Allyn and Bacon, 1995.

Riesch, R., L. G. Barrett-Lennard, G. M. Ellis, J. K. B. Ford, and V. B. Deecke. "Cultural Traditions and the Evolution of Reproductive Isolation: Ecological Speciation in Killer Whales?" *Biological Journal of the Linnean Society* 106 (2012): 1–17.

Rincover, A., and J. Devany. "The Application of Sensory Extinction Procedures to Self-Injury." *Analysis and Intervention in Developmental Disabilities* 2 (1982): 67–81.

Roberts, B. "Notes on the Birds of Central and South-east Iceland with Special Reference to Food Habits." *Ibis* 13 (1934): 239–64.

Roberts, C., M. T. Harvey, M. E. May, M. G. Valdovinos, T. G. Patterson, M. H. Couppis, and C. H. Kennedy. "Varied Effects of Conventional Antiepileptics on Responding Maintained by Negative versus Positive Reinforcement." *Physiology & Behavior* 93 (2008): 612–21.

Robinson, P. W., D. F. Foster, and C. V. Bridges. "Errorless Learning in Newborn Chicks." *Animal Learning & Behavior* 4 (1976): 266–68.

Roderick, C., M. Pitchford, and A. Miller. "Reducing Aggressive Playground Behaviour by Means of a School-Wide 'Raffle.'" *Educational Psychology in Practice* 13 (1997): 57–63.

Rodin, J., and E. J. Langer. "Long-Term Effects of a Control-Relevant Intervention with the Institutionalized Aged." *Journal of Personality and Social Psychology* 35 (1977): 897–902.

Rogers, S. J., and L. A. Vismara. "Evidence-Based Comprehensive Treatments for Early Autism." *Journal of Clinical Child and Adolescent Psychology* 37 (2008): 8–38.

Roll, J. M., S. T. Higgins, and G. J. Badger. "An Experimental Comparison of Three Different Schedules of Reinforcement of Drug Abstinence using Cigarette Smoking as an Exemplar." *Journal of Applied Behavior Analysis* 29 (1996): 495–504.

Rosenthal R., and L. Jacobson. *Pygmalion in the Classroom: Teacher Expectation and Pupils' Intellectual Development.* New York: Irvington, 1968. Expanded edition published in 1992.

Rossi, A. P., and C. Ades. "A Dog at the Keyboard: Using Arbitrary Signs to Communicate Requests." *Animal Cognition* 11 (2008): 329–38.

Roth, T. L., F. D. Lubin, A. J. Funk, and J. D. Sweatt. "Lasting Epigenetic Influence of Early-Life Adversity on the *BDNF* Gene." *Biological Psychiatry* 65 (2009): 760–69.

Rothstein, J. B., G. Jensen, and A. Neuringer. "Human Choice among Five Alternatives When Reinforcers Decay." *Behavioural Processes* 78 (2008): 231–39.

Rounding, V. *Catherine the Great.* New York: St. Martin's Press, 2006.

Routh, D. K. "Conditioning of Vocal Response Differentiation in Infants." *Developmental Psychology* 1 (1969): 219–26.

Russell, M., K. A. Dark, R. W. Cummins, G. Ellman, E. Callaway, and H. V. Peeke. "Learned Histamine Release." *Science* 225 (1984): 733–34.

Ruxton, G. D., and M. H. Hansell. "Fishing with a Bait or Lure: A Brief Review of the Cognitive Issues." *Ethology* 117 (2011): 1–9.

Safriel, U. N., B. J. Ens, and A. Kaiser. "Rearing to Independence." In *The Oystercatcher: Individuals to Populations*, edited by J. D. Goss-Custard, pp. 210–50. Oxford: Oxford University Press, 1996.

Sagan, C., and A. Druyan. *Shadows of Forgotten Ancestors: A Search for Who We Are.* New York: Random House, 1993.

Saigh, P. A., and A. M. Umar. "The Effects of a Good Behavior Game on the Disruptive Behavior of Sudanese Elementary School Students." *Journal of Applied Behavior Analysis* 16 (1983): 339–44.

Salzinger, K., S. J. Freimark, S. P. Fairhurst, and F. D. Wolkoff. "Conditioned Reinforcement in the Goldfish." *Science* 160 (1968): 1471–72.

Savage, T. "Shaping: A Multiple Contingencies Analysis and Its Relevance to Behaviour-Based Robotics." *Connection Science* 13 (2001): 199–234.

Savage-Rumbaugh, E. S. "Verbal Behavior at a Procedural Level in the Chimpanzee." *Journal of the Experimental Analysis of Behavior* 41 (1984): 223–50.

Savage-Rumbaugh, E. S., D. M. Rumbaugh, and S. Boysen. "Symbolic Communication between Two Chimpanzees (*Pan troglodytes*)." *Science* 201 (1978): 641–44.

Savastano, G., A. Hanson, and C. M. Savastano. "The Development of an Operant Conditioning Training Program for New World Primates at the Bronx Zoo." *Journal of Applied Animal Welfare Science* 6 (2003): 247–61.

Scalera, B., and M. Bavieri. "Role of Conditioned Taste Aversion on the Side Effects of Chemotherapy in Cancer Patients." In *Conditioned Taste Aversion: Behavioral and Neural Processes*, edited by S. Reilly and T. R. Schachtman, pp. 513–41. Oxford: Oxford University Press, 2009.

Schick, K. D., N. Toth, G. Garufi, E. S. Savage-Rumbaugh, D. Rumbaugh, and R. Sevcik. "Continuing Investigations into the Stone Tool-Making and Tool-Using Capabilities of a Bonobo (*Pan paniscus*)." *Journal of Archaeological Science* 26 (1999): 821–32.

Schiff, N. B., and I. M. Ventry. "Communication Problems in Hearing Children of Deaf Parents." *Journal of Speech and Hearing Disorders* 41 (1976): 348–58.

Schlinger, H., E. Blakeley, and T. Kaczor. "Pausing under Variable-Ratio Schedules: Interaction of Reinforcer Magnitude, Variable-Ratio Size, and Lowest Ratio." *Journal of the Experimental Analysis of Behavior* 53 (1990): 133–39.

Schlosser, E. *Fast Food Nation: The Dark Side of the All-American Meal.* Boston: Houghton Mifflin, 2001.

Schlund, M. W., S. Magee, and C. D. Hudgins. "Human Avoidance and Approach Learning: Evidence for Overlapping Neural Systems and Experiential Avoidance Modulation of Avoidance Neurocircuitry." *Behavioural Brain Research* 225 (2011): 437–48.

Schneider, S. M. "Rats' Behavior in Two Different Home Cages." *Humane Innovations and Alternatives* 2 (1988): 39–42.

———. "The Role of Contiguity in Free-Operant Unsignaled Delay of Positive Reinforcement: A Brief Review." *Psychological Record* 40 (1990): 239–57.

———. "A Two-Stage Model for Concurrent Sequences." *Behavioural Processes* 78 (2008): 429–41.

Schneider, S. M., and M. Davison. "Demarcated Response Sequences and Generalised Matching." *Behavioural Processes* 70 (2005): 51–61.

———. "Molecular Order in Concurrent Response Sequences." *Behavioural Processes* 73 (2006): 187–98.

Schneider, S. M., and R. Lickliter. "Choice in Quail Neonates: The Origins of Generalized Matching." *Journal of the Experimental Analysis of Behavior* 94 (2010a): 315–26.

———. "Operant Generalization in Quail Neonates after Intradimensional Training: Distinguishing Positive and Negative Reinforcement." *Behavioural Processes* 83 (2010b): 1–7.

Schneider, S. M., and E. K. Morris. "Sequences of Spaced Responses: Behavioral Units and the Role of Contiguity." *Journal of the Experimental Analysis of Behavior* 58 (1992): 537–55.

Schroedel, J. R. *Alone in a Crowd: Women in the Trades Tell Their Stories.* Philadelphia: Temple University Press, 1985.

Schulz, L. O., P. H. Bennett, E. Ravussin, J. R. Kidd, K. K. Kidd, J. Esparza, and M. E. Valencia. "Effects of Traditional and Western Environments on Prevalence of Type 2 Diabetes in Pima Indians in Mexico and the U.S." *Diabetes Care* 29 (2006): 1866–71.

Schuster, J. C. "Acoustic Signals of Passalid Beetles: Complex Repertoires." *Florida Entomologist* 66 (1983): 486–96.

Schusterman, R. J. "Vocal Learning in Mammals with Special Emphasis on Pinnipeds." In *Evolution of Communicative Flexibility: Complexity, Creativity, and Adapt-*

ability in Human and Animal Communication, edited by D. K. Oller and U. Griebel, pp. 41–70. Cambridge, MA: MIT Press, 2008.

Schweitzer, J. B., and B. Sulzer-Azaroff. "Self-Control: Teaching Tolerance for Delay in Impulsive Children." *Journal of the Experimental Analysis of Behavior* 50 (1988): 173–86.

———. "Self-Control in Boys with Attention Deficit Hyperactivity Disorder: Effects of Added Stimulation and Time." *Journal of Child Psychology and Psychiatry* 36 (1995): 671–86.

Scobie, S. R., and D. C. Gold. "Differential Reinforcement of Low Rates in Goldfish." *Learning & Behavior* 3 (1975): 143–46.

Seligman, M. E., and S. F. Maier. "Failure to Escape Traumatic Shock." *Journal of Experimental Psychology* 74 (1967): 1–9.

Serling, R. *Patterns*. New York: Simon and Schuster, 1957.

Sethi-Iyengar, S., G. Huberman, and W. Jiang. "How Much Choice Is Too Much? Contributions to 401(k) Retirement Plans." In *Pension Design and Structure: New Lessons from Behavioral Finance*, edited by O. S. Mitchell and S. P. Utkus, pp. 83–95. Oxford: Oxford University Press, 2004.

Seyfarth, R. M., and D. L. Cheney. "Vocal Development in Vervet Monkeys." *Animal Behaviour* 34 (1986): 1640–58.

Shapira, A., and M. C. Madsen. "Cooperative and Competitive Behavior of Kibbutz and Urban Children in Israel." *Child Development* 40 (1969): 609–17.

Sharma, J., A. Angelucci, and M. Sur. "Induction of Visual Orientation Modules in Auditory Cortex." *Nature* 404 (2000): 841–47.

Sharp, W. G., D. L. Jaquess, J. F. Morton, and C. V. Herzinger. "Pediatric Feeding Disorders: A Quantitative Synthesis of Treatment Outcomes." *Clinical Child and Family Psychology Review* 13 (2010): 348–65.

Shead, N. W., and D. C. Hodgins. "Probability Discounting of Gains and Losses: Implications for Risk Attitudes and Impulsivity." *Journal of the Experimental Analysis of Behavior* 92 (2009): 1–16.

Shenk, D. *The Genius in All of Us: Why Everything You've Been Told about Genetics, Talent, and IQ Is Wrong*. New York: Doubleday, 2010.

Shepherdson, D. J., K. Carlstead, J. D. Mellen, and J. Seidensticker. "The Influence of Food Presentation on the Behavior of Small Cats in Confined Environments." *Zoo Biology* 12 (1993): 203–16.

Sherif, M., and C. W. Sherif. "Ingroup and Intergroup Relations." In *Introduction to Psychology*, edited by J. O. Whittaker, pp. 541–73. Philadelphia: Saunders, 1965.

Sherman, J. G. "Reminiscences: Excerpts from the Diary of a Behaviorist: The Search for a Science of Psychology, the Early Years." *Behavior and Social Issues* 5 (1995): 51–71.

Shigemitsu, Y. "Different Interpretations of Pauses in Natural Conversation—Japanese, Chinese and Americans." *Academic Reports of the Faculty of Engineering, Tokyo Polytechnic University* 28 (2005): 8–14.

Shizgal, P. "On the Neural Computation of Utility: Implications from Studies of Brain Stimulation Reward." In *Well-Being: The Foundations of Hedonic Psychology*, edited by D. Kahneman, E. Diener, and N. Schwarz, pp. 502–26. New York: Russell Sage, 1999.

Shoda, Y., W. Mischel, and P. K. Peake. "Predicting Adolescent Cognitive and Self-Regulatory Competencies from Preschool Delay of Gratification: Identifying Diagnostic Conditions." *Developmental Psychology* 26 (1990): 978–86.

Shyne A., and M. Block. "The Effects of Husbandry Training on Stereotypic Pacing in Captive African Wild Dogs (*Lycaon pictus*)." *Journal of Applied Animal Welfare Science* 13 (2010): 56–65.

Siegel, E., and H. Rachlin. "Soft Commitment: Self-Control Achieved by Response Persistence." *Journal of the Experimental Analysis of Behavior* 64 (1995): 117–28.

Siegel, S., and D. W. Ellsworth. "Pavlovian Conditioning and Death from Apparent Overdose of Medically Prescribed Morphine: A Case Report." *Bulletin of the Psychonomic Society* 24 (1986): 278–80.

Siegel, S., and B. M. C. Ramos. "Applying Laboratory Research: Drug Anticipation and the Treatment of Drug Addiction." *Experimental and Clinical Psychopharmacology* 10 (2002): 162–83.

Silberberg, A., J. R. Thomas, and N. Berendzen. "Human Choice on Concurrent Variable-Interval Variable-Ratio Schedules." *Journal of the Experimental Analysis of Behavior* 56 (1991): 575–84.

Silva, M. T. A., F. L. Gonçalves, and M. Garcia-Mijares. "Neural Events in the Reinforcement Contingency." *Behavior Analyst* 30 (2007): 17–30.

Simon, N. W., I. A. Mendez, and B. Setlow. "Cocaine Exposure Causes Long-Term Increases in Impulsive Choice." *Behavioral Neuroscience* 121 (2007): 543–49.

Simons, D. J., and C. F. Chabris. "Gorillas in Our Midst: Sustained Inattentional Blindness for Dynamic Events." *Perception* 28 (1999): 1059–74.

Singer, T., B. Seymour, J. O'Doherty, H. Kaube, R. J. Dolan, and C. D. Frith. "Empathy for Pain Involves the Affective but Not the Sensory Components of Pain." *Science* 303 (2004): 1157–62.

Singh, B. R. "Teaching Methods for Reducing Prejudice and Enhancing Academic Achievement for All Children." *Educational Studies* 17 (1991): 157–71.

Siqueland, E. R., and L. P. Lipsitt. "Conditioned Head-Turning in Human New-Borns." *Journal of Experimental Child Psychology* 3 (1966): 356–76.

Skinner, B. F. "Behaviorism at Fifty." *Science* 140 (1963): 951–58.

———. *The Behavior of Organisms*. Englewood Cliffs, NJ: Prentice-Hall, 1938.

———. "The Evolution of Verbal Behavior." *Journal of the Experimental Analysis of Behavior* 45 (1986): 115–22.

———. "How to Discover What You Have to Say—A Talk to Students." *Behavior Analyst* 4 (1981a): 1–7.

———. *A Matter of Consequences: Part Three of an Autobiography.* New York: Knopf, 1983.

———. *Notebooks.* Edited by R. Epstein. Englewood Cliffs, NJ: Prentice-Hall, 1980.

———. "Pigeons in a Pelican." *American Psychologist* 15 (1960): 28–37.

———. *Science and Human Behavior.* New York: Free Press, 1957a.

———. "Selection by Consequences." *Science* 213 (1981b): 501–504.

———. *The Shaping of a Behaviorist.* New York: Knopf, 1979.

———. *The Technology of Teaching.* Englewood Cliffs, NJ: Prentice-Hall, 1968.

———. *Verbal Behavior.* New York: Prentice-Hall, 1957b.

———. *Walden Two.* New York: Macmillan, 1948.

Slotnick, B., L. Hanford, and W. Hodos. "Can Rats Acquire an Olfactory Learning Set?" *Journal of Experimental Psychology: Animal Behavior Processes* 26 (2000): 399–415.

Sloutsky, V. M., and A. V. Fisher. "Attentional Learning and Flexible Induction: How Mundane Mechanisms Give Rise to Smart Behaviors." *Child Development* 79 (2008): 639–51.

Small, D., G. Loewenstein, and P. Slovic. "Sympathy and Callousness: The Impact of Deliberative Thought on Donations to Identifiable and Statistical Victims." *Organizational Behavior and Human Decision Processes* 102 (2007): 143–53.

Snyder, J. J., and G. R. Patterson. "Individual Differences in Social Aggression: A Test of a Reinforcement Model of Socialization in the Natural Environment." *Behavior Therapy* 26 (1995): 371–91.

Sonoda, H., S. Kohnoe, T. Yamazato, Y. Satoh, G. Morizono, K. Shikata, M. Morita, et al. "Colorectal Cancer Screening with Odour Material by Canine Scent Detection." *Gut* 60 (2011): 814–19.

Spiga, R., S. Maxwell, R. A. Meisch, and J. Grabowski. "Human Methadone Self-Administration and the Generalized Matching Law." *Psychological Record* 55 (2005): 525–38.

Staats, A. W. *Learning, Language, and Cognition.* New York: Holt, Rinehart and Winston, 1968.

Staats, A. W., and C. K. Staats. "Attitudes Established by Classical Conditioning." *Journal of Abnormal and Social Psychology* 57 (1958): 37–40.

Stackhouse, K. "Interview with a Bug-Eater." *Sierra* 96 (March 2011): 36–37.

Stebbins, K. R. "Going Like Gangbusters: Transnational Tobacco Companies 'Making a Killing' in South America." *Medical Anthropology Quarterly* 15 (2001): 147–69.

Stein, L., B. G. Xue, and J. D. Belluzzi. "A Cellular Analogue of Operant Conditioning." *Journal of the Experimental Analysis of Behavior* 60 (1993): 41–53.

Steinbeck, J. *Travels with Charley: In Search of America*. New York: Penguin, 1962.

Steiner, S. S., B. Beer, and M. Shaffer. "Escape from Self-Produced Rates of Rewarding Brain Stimulation." *Science* 163 (1969): 90–91.

Stewart, T. L., J. R. Laduke, C. Bracht, B. A. M. Sweet, and K. E. Gamarel. "Do the 'Eyes' Have It? A Program Evaluation of Jane Elliott's 'Blue-Eyes/Brown-Eyes' Diversity Training Exercise." *Journal of Applied Social Psychology* 33 (2003): 1898–1921.

Stitzer, M. L., and R. Vandrey. "Contingency Management: Utility in the Treatment of Drug Abuse Disorders." *Clinical Pharmacology and Therapeutics* 83 (2008): 644–47.

Stroud, C. B., J. Davila, C. Hammen, and S. Vrshek-Schallhorn. "Severe and Nonsevere Events in First Onsets versus Recurrences of Depression: Evidence for Stress Sensitization." *Journal of Abnormal Psychology* 120 (2011): 142–54.

Sullivan, R. M., and M. Leon. "One-Trial Olfactory Learning Enhances Olfactory Bulb Responses to an Appetitive Conditioned Odor in 7-Day-Old Rats." *Developmental Brain Research* 35 (1987): 307–11.

Sulzer-Azaroff, B., and J. Austin. "Does BBS work? Behavior-Based Safety and Injury Reduction: A Survey of the Evidence." *Professional Safety* (2000): 19–24.

Sumpter, C. E., T. M. Foster, and W. Temple. "Assessing Animals' Preferences: Concurrent Schedules of Reinforcement." *International Journal of Comparative Psychology* 15 (2002): 107–26.

Sumpter, C. E., W. Temple, and T. M. Foster. "Response Form, Force, and Number: Effects on Concurrent-Schedule Performance." *Journal of the Experimental Analysis of Behavior* 70 (1998): 45–68.

Suomi, S. J. "How Gene-Environment Interactions Can Shape the Development of Socioemotional Regulation in Rhesus Monkeys." In *Emotional Regulation and Developmental Health: Infancy and Early Childhood*, edited by B. S. Zuckerman, A. F. Zuckerman, and N. A. Fox, pp. 5–26. New Brunswick, NJ: Johnson and Johnson Pediatric Institute, 2002.

Sutherland, A. *What Shamu Taught Me about Life, Love, and Marriage: Lessons for People from Animals and Their Trainers*. New York: Random House, 2008.

Svartdal, F. "Operant Modulation of Low-Level Attributes of Rule-Governed Behavior by Nonverbal Contingencies." *Learning and Motivation* 22 (1991): 406–20.

———. "Sensitivity to Nonverbal Operant Contingencies: Do Limited Processing Resources Affect Operant Conditioning in Humans?" *Learning and Motivation* 23 (1992): 383–405.

Swann, P. G. "Heterochromia: Signs and Symptoms." *Optometry Today* 39 (1999): 30–32.

Swengel, J. "Birds Watching People." *Bird Watcher's Digest* 24, March 2001, pp. 63–67.

Taglialatela, J. P., S. Savage-Rumbaugh, and L. A. Baker. "Vocal Production by a Language-Competent *Pan paniscus*." *International Journal of Primatology* 24 (2003): 1–17.

Tamashiro, K. L. K., T. Wakayama, H. Akutsu, Y. Yamazaki, J. L. Lachey, M. D. Wortman, R. J. Seeley, et al. "Cloned Mice Have an Obese Phenotype Not Transmitted to Their Offspring." *Nature: Medicine* 8 (2002): 262–67.

Tang, A. C. "Neonatal Exposure to Novel Environment Enhances Hippocampal-Dependent Memory Function during Infancy and Adulthood." *Learning and Memory* 8 (2001): 257–64.

Tanol, G., L. Johnson, J. McComas, and E. Cote. "Responding to Rule Violations or Rule Following: A Comparison of Two Versions of the Good Behavior Game with Kindergarten Students." *Journal of School Psychology* 48 (2010): 337–55.

Tarou, L. R., and M. J. Bashaw. "Maximizing the Effectiveness of Environmental Enrichment: Suggestions from the Experimental Analysis of Behavior." *Applied Animal Behaviour Science* 102 (2007): 189–204.

Tatham, T. A., and B. A. Wanchisen. "Behavioral History: A Definition and Some Common Findings from Two Areas of Research." *Behavior Analyst* 21 (1998): 241–51.

Taub, E., G. Uswatte, D. K. King, D. Morris, J. E. Crago, and A. Chatterjee. "A Placebo-Controlled Trial of Constraint-Induced Movement Therapy for Upper Extremity after Stroke." *Stroke* 37 (2006): 1045–49.

Tavris, C. *Anger: The Misunderstood Emotion*. Rev. ed. New York: Simon and Schuster, 1989.

———. "The Frozen World of the Familiar Stranger: An Interview with Stanley Milgram." *Psychology Today* 8 (1974): 71–73, 76–78, 80.

Tavris, C., and E. Aronson. *Mistakes Were Made (but Not by Me): Why We Justify Foolish Beliefs, Bad Decisions, and Hurtful Acts*. Orlando, FL: Harcourt, 2007.

Taylor, B. A., and H. Hoch. "Teaching Children with Autism to Respond to and Initiate Bids for Joint Attention." *Journal of Applied Behavior Analysis* 41 (2008): 377–91.

Taylor, T. K., and A. Biglan. "Behavioral Family Interventions for Improving Child-Rearing: A Review of the Literature for Clinicians and Policy Makers." *Clinical Child and Family Psychology Review* 1 (1998): 41–60.

Teale, E. W. *Days without Time: Adventures of a Naturalist*. New York: Dodd, Mead, 1948.

———. *A Walk through the Year*. New York: Dodd, Mead, 1978.

ten Cate, C. "Perceptual Mechanisms in Imprinting and Song Learning." In *Causal*

Mechanisms of Behavioural Development, edited by J. A. Hogan and J. J. Bolhuis, pp. 116–46. Cambridge: Cambridge University Press, 1994.

Thaler, R. H., and S. Benartzi. "Save More Tomorrow: Using Behavioral Economics to Increase Employee Saving." *Journal of Political Economy* 112 (2004): 164–87.

Thaler, R. H., and C. R. Sunstein. *Nudge: Improving Decisions about Health, Wealth, and Happiness.* New Haven, CT: Yale University Press, 2008.

Thomas, D. R., W. C. Becker, and M. Armstrong. "Production and Elimination of Disruptive Classroom Behavior by Systematically Varying Teacher's Behavior." *Journal of Applied Behavior Analysis* 1 (1968): 35–45.

Thompson, R. H., and B. A. Iwata. "A Descriptive Analysis of Social Consequences Following Problem Behavior." *Journal of Applied Behavior Analysis* 34 (2001): 169–78.

Thornton, A., and K. McAuliffe. "Teaching in Wild Meerkats." *Science* 313 (2006): 227–29.

Till, B. D., S. M. Stanley, and R. Priluck. "Classical Conditioning and Celebrity Endorsers: An Examination of Belongingness and Resistance to Extinction." *Psychology and Marketing* 25 (2008): 179–96.

Tingstrom, D. H., H. E. Sterling-Turner, and S. M. Wilczynski. "The Good Behavior Game: 1969–2002." *Behavior Modification* 30 (2006): 225–53.

Tinker, J. E., and J. A. Tucker. "Motivations for Weight Loss and Behavior Change Strategies Associated with Natural Recovery from Obesity." *Psychology of Addictive Behaviors* 11 (1997): 98–106.

Tomasello, M. "Do Young Children Have Adult Syntactic Competence?" *Cognition* 74 (2000): 209–53.

Tumer, E. C., and M. S. Brainard. "Performance Variability Enables Adaptive Plasticity of 'Crystallized' Adult Birdsong." *Nature* 450 (2007): 1240–44.

Turnbull, C. M. "The Politics of Non-Aggression." In *Learning Non-Aggression: The Experience of Non-Literate Societies*, edited by A. Montagu, pp. 161–221. New York: Oxford University Press, 1978.

Udell, M. A. R., N. R. Dorey, and C. D. L. Wynne. "Wolves Outperform Dogs in Following Human Social Cues." *Animal Behaviour* 76 (2008): 1767–73.

Ulrich, R. "Pain as a Cause of Aggression." *American Zoologist* 6 (1966): 643–62.

Valsecchi, A. M., D. Mainardi, and M. Mainardi. "Individual and Social Experiences in the Establishment of Food Preferences in Mice." In *Behavioral Aspects of Feeding*, edited by B. G. Galef, M. Mainardi, and A. Valsecchi, pp. 103–24. Boca Raton, FL: CRC Press, 1994.

Van Driesche, J., and R. Van Driesche. *Nature out of Place: Biological Invasions in the Global Age.* Washington, DC: Island Press, 2000.

Van Egeren, L. A., M. S. Barratt, and M. A. Roach. "Mother-Infant Responsiveness: Timing, Mutual Regulation, and Interactional Context." *Developmental Psychology* 37 (2001): 684–97.

Van Haaren, F., A. Van Hest, and N. E. Van De Poll. "Self-Control in Male and Female Rats." *Journal of the Experimental Analysis of Behavior* 49 (1988): 201–11.

Van Houten, R., and P. A. Nau. "A Comparison of the Effects of Fixed and Variable Ratio Schedules of Reinforcement on the Behavior of Deaf Children." *Journal of Applied Behavior Analysis* 13 (1980): 13–21.

Van Schaik, C. P., M. Ancrenaz, G. Borgen, B. Galdikas, C. D. Knott, I. Singleton, A. Suzuki, S. S. Utami, and M. Merrill. "Orangutan Cultures and the Evolution of Material Culture." *Science* 299 (2003): 102–105.

Vargha-Khadem, F., L. Carr, E. Isaacs, E. Brett, C. Adams, and M. Mishkin. "Onset of Speech after Left Hemispherectomy in a Nine-Year-Old Boy." *Brain* 120 (1997): 159–82.

Vasconcelos, M. "Transitive Inference in Non-Human Animals: An Empirical and Theoretical Analysis." *Behavioural Processes* 78 (2008): 313–34.

Vaughan, W. "Formation of Equivalence Sets in Pigeons." *Journal of Experimental Psychology: Animal Behavior Processes* 14 (1988): 36–42.

Veblen, T. *The Theory of the Leisure Class*. Mineola, NY: Dover, 1994. Originally published in 1899.

Vollaro, D. R. "Lincoln, Stowe, and the 'Little Woman/Great War' Story: The Making, and Breaking, of a Great American Anecdote." *Journal of the Abraham Lincoln Association* 30 (2009): 18–34.

Vollmer, T. R., and B. A. Iwata. "Establishing Operations and Reinforcement Effects." *Journal of Applied Behavior Analysis* 24 (1991): 279–91.

Volpp, K. G., A. B. Troxel, M. V. Pauly, H. A. Glick, A. Puig, D. A. Asch, R. Galvin, et al. "A Randomized, Controlled Trial of Financial Incentives for Smoking Cessation." *New England Journal of Medicine* 360 (2009): 699–709.

Vyse, S. A. *Believing in Magic: The Psychology of Superstition*. Oxford: Oxford University Press, 1997.

Wager, T. D., J. K. Rilling, E. E. Smith, A. Sokolik, K. L. Casey, R. J. Davidson, S. M. Kosslyn, R. M. Rose, and J. D. Cohen. "Placebo-Induced Changes in fMRI in the Anticipation and Experience of Pain." *Science* 303 (2004): 1162–67.

Wagner, G. A., and E. K. Morris. "'Superstitious' Behavior in Children." *Psychological Record* 37 (1987): 471–88.

Walker, H. M. *The Acting-Out Child: Coping with Classroom Disruption*. Boston: Allyn and Bacon, 1979.

Walker, H. M., R. H. Mattson, and N. K. Buckley. "The Functional Analysis of Behavior within an Experimental Class Setting." In *An Empirical Basis for Change in Education: Selections on Behavioral Psychology for Teachers*, edited by W. C. Becker, pp. 236–63. Chicago: Science Research Associates, 1971.

Wallace, I. "Self-Control Techniques of Famous Novelists." *Journal of Applied Behavior Analysis* 10 (1977): 515–25.

Wallace, M. D., B. A. Iwata, and G. P. Hanley. "Establishment of Mands Following Tact Training as a Function of Reinforcer Strength." *Journal of Applied Behavior Analysis* 39 (2006): 17–24.

Walpole, C. W., E. M. Roscoe, and W. V. Dube. "Use of a Differential Observing Response to Expand Restricted Stimulus Control." *Journal of Applied Behavior Analysis* 40 (2007): 707–12.

Walsh, K., D. Glaser, and D. D. Wilcox. *What Education Schools Aren't Teaching—And What Elementary Teachers Aren't Learning.* Washington, DC: National Council on Teacher Quality, May 2006.

Wanchisen, B. A., T. A. Tatham, and S. E. Mooney. "Variable-Ratio Conditioning History Produces High- and Low-Rate Fixed-Interval Performance in Rats." *Journal of the Experimental Analysis of Behavior* 52 (1989): 167–79.

Ward, J. "Variation of Reinforcement in Performance of a Motor Skill." *Perceptual and Motor Skills* 43 (1976): 149–50.

Watanabe, H., and M. Mizunami. "Pavlov's Cockroach: Classical Conditioning of Salivation in an Insect." *PLoS ONE* 2 (2007): e529.

Watanabe, S., and M. Nemoto. "Reinforcing Property of Music in Java Sparrows (*Padda oryzivora*)." *Behavioural Processes* 43 (1998): 211–18.

Watanabe, S., J. Sakamoto, and M. Wakita. "Pigeons' Discrimination of Paintings by Monet and Picasso." *Journal of the Experimental Analysis of Behavior* 63 (1995): 165–74.

Waterland, R. A., and R. L. Jirtle. "Transposable Elements: Targets for Early Nutritional Effects on Epigenetic Gene Regulation." *Molecular and Cellular Biology* 23 (2003): 5293–300.

Watkins, C. L. *Project Follow Through: A Case Study of Contingencies Influencing Instructional Practices of the Educational Establishment.* Cambridge, MA: Cambridge Center for Behavioral Studies, 1997.

Watson, T. S., and H. E. Sterling. "Brief Functional Analysis and Treatment of a Vocal Tic." *Journal of Applied Behavior Analysis* 31 (1998): 471–74.

Weaver, I. C. G., N. Cervoni, F. A. Champagne, A. C. D'Alessio, S. Sharma, J. R. Seck, S. Dymov, M. Szyf, and M. J. Meaney. "Epigenetic Programming by Maternal Behavior." *Nature Neuroscience* 7 (2004): 847–54.

Webster-Stratton, C., M. J. Reid, and M. Hammond. "Treating Children with Early-Onset Conduct Problems: Intervention Outcomes for Parent, Child, and Teacher Training." *Journal of Clinical Child and Adolescent Psychology* 33 (2004): 105–24.

Weiner, H. "Conditioning History and Human Fixed-Interval Performance." *Journal of the Experimental Analysis of Behavior* 7 (1964): 383–85.

Weiss, M. R. "Innate Colour Preferences and Flexible Colour Learning in the Pipevine Swallowtail." *Animal Behaviour* 53 (1997): 1043–52.

Weiss, M. R., E. E. Wilson, and I. Castellanos. "Predatory Wasps Learn to Overcome the Shelter Defences of Their Larval Prey." *Animal Behaviour* 68 (2004): 45–54.

Wells, P. H. "Training Flatworms in a Van Oye Maze." In *Chemistry of Learning*, edited by W. Corning and S. Ratner, pp. 251–54. New York: Plenum, 1967.

Werner, T. K., and T. W. Sherry. "Behavioral Feeding Specialization in *Pinaroloxias inornata*, the Darwin's Finch of Cocos Island, Costa Rica." *Proceedings of the National Academy of Sciences* 84 (1987): 5506–10.

Whalen, C., and L. Schreibman. "Joint Attention Training for Children with Autism Using Behavior Modification Procedures." *Journal of Child Psychology and Psychiatry* 44 (2003): 456–68.

White, K. D. "Salivation: The Significance of Imagery in Its Voluntary Control." *Psychophysiology* 15 (1978): 196–203.

Whitehurst G. J., and M. C. Valdez-Menchaca. "What Is the Role of Reinforcement in Early Language Acquisition?" *Child Development* 59 (1988): 430–40.

Wich, S. A., K. B. Swartz, M. E. Hardus, A. R. Lameira, E. Stromberg, and R. W. Shumaker. "A Case of Spontaneous Acquisition of a Human Sound by an Orangutan." *Primates* 50 (2009): 56–64.

Wilder, D. A. "Some Determinants of the Persuasive Power of In-Groups and Out-Groups: Organization of Information and Attribution of Independence." *Journal of Personality and Social Psychology* 59 (1990): 1202–13.

Wilkinson, R., and K. Pickett. *The Spirit Level: Why Greater Equality Makes Societies Stronger*. New York: Bloomsbury, 2009.

Williams, W. M. "Are We Raising Smarter Children Today? School- and Home-Related Influences on IQ." In *The Rising Curve: Long-Term Gains in IQ and Related Measures*, edited by U. Neisser, pp. 125–54. Washington, DC: American Psychological Association, 1998.

Wilsoncroft, W. E. "Babies by Bar-Press: Maternal Behavior in the Rat." *Behavior Research Methods* 1 (1969): 229–30.

Winter, A., and J. E. Hillerton. "Behaviour Associated with Feeding and Milking of Early Lactation Cows Housed in an Experimental Automatic Milking System." *Applied Animal Behaviour Science* 46 (1995): 1–15.

Wise, R. A. "Dopamine, Learning and Motivation." *Nature Reviews Neuroscience* 5 (2004): 1–12.

Wodehouse, P. G. *A Few Quick Ones*. New York: Simon and Schuster, 1959.

Wolf, M. M., T. R. Risley, and H. Mees. "Application of Operant Conditioning Procedures to the Behaviour Problems of an Autistic Child." *Behaviour Research and Therapy* 1 (1964): 305–12.

Wolf, S. L., P. A. Thompson, C. J. Winstein, J. P. Miller, S. R. Blanton, D. S. Nichols-Larsen, D. M. Morris, et al. "The EXCITE Stroke Trial: Comparing

Early and Delayed Constraint-Induced Movement Therapy." *Stroke* 41 (2010): 2309–15.

Wolf, S. L., C. J. Winstein, J. P. Miller, P. A. Thompson, E. Taub, G. Uswatte, D. Morris, S. Blanton, D. Nichols-Larsen, and P. C. Clark. "Retention of Upper Limb Function in Stroke Survivors Who Have Received Constraint-Induced Movement Therapy: The EXCITE Randomised Trial." *Lancet Neurology* 7 (2008): 33–40.

Wolitzky-Taylor, K. B., J. D. Horowitz, M. B. Powers, and M. J. Telch. "Psychological Approaches in the Treatment of Specific Phobias: A Meta-Analysis." *Clinical Psychological Review* 28 (2008): 1021–37.

Woods, S. C., and D. S. Ramsay. "Pavlovian Influences over Food and Drug Intake." *Behavioral Brain Research* 110 (2000): 175–82.

Woods, S. C., J. R. Vasselli, E. Kaestner, G. A. Szakmary, P. Milburn, and M. V. Vitiello. "Conditioned Insulin Secretion and Meal Feeding in Rats." *Journal of Comparative and Physiological Psychology* 91 (1977): 128–33.

Woolfolk, M. E., W. Castellan, and C. I. Brooks. "Pepsi versus Coke: Labels, Not Tastes, Prevail." *Psychological Reports* 52 (1983): 185–86.

Wright, A. A., R. G. Cook, and D. F. Kendrick. "Relational and Absolute Stimulus Learning by Monkeys in a Memory Task." *Journal of the Experimental Analysis of Behavior* 52 (1989): 237–48.

Wunderle, J. M. "Age-Specific Foraging Proficiency in Birds." *Current Ornithology* 8 (1991): 273–324.

Wynne, C. D. L. "Pigeon Transitive Inference: Tests of Simple Accounts of a Complex Performance." *Behavioural Processes* 39 (1997): 95–112.

Xi, Z.-X., and E. A. Stein. "Baclofen Inhibits Heroin Self-Administration Behavior and Mesolimbic Dopamine Release." *Journal of Pharmacology and Experimental Therapeutics* 290 (1999): 1369–74.

Yin, J. C. P., J. S. Wallach, M. Del Vecchio, E. L. Wilder, H. Zhou, W. G. Quinn, and T. Tully. "Induction of a Dominant Negative CREB Transgene Specifically Blocks Long-Term Memory Formation in *Drosophila*." *Cell* 79 (1994): 49–58.

Young, J. M., P. J. Krantz, L. E. McClannahan, and C. L. Poulson. "Generalized Imitation and Response-Class Formation in Children with Autism." *Journal of Applied Behavior Analysis* 27 (1994): 685–97.

Youngentob, S. L., and J. I. Glendinning. "Fetal Ethanol Exposure Increases Ethanol Intake by Making It Smell and Taste Better." *Proceedings of the National Academy of Sciences* 106 (2009): 5359–64.

Youngentob, S. L., J. C. Molina, N. E. Spear, and L. M. Youngentob. "The Effect of Gestational Ethanol Exposure on Voluntary Ethanol Intake in Early Postnatal and Adult Rats." *Behavioral Neuroscience* 121 (2007): 1306–15.

Zaloga, S. *The Red Army of the Great Patriotic War, 1941–5.* Oxford: Osprey, 1989.

Zeilberger, J., S. E. Sampen, and H. N. Sloane. "Modification of a Child's Problem Behaviors in the Home with the Mother as Therapist." *Journal of Applied Behavior Analysis* 1 (1968): 47–53.

Zentall, T. R., E. A. Wasserman, O. F. Lazareva, R. K. R. Thompson, and M. J. Rattermann. "Concept Learning in Animals." *Comparative Cognition & Behavior Reviews* 3 (2008): 13–45.

Zickefoose, J. "After the Spark." *Bird Watcher's Digest* 23, September 2000, pp. 18–23.

Zimmer, C. *Microcosm: E. coli and the New Science of Life.* New York: Pantheon, 2008.

Zimmerman, D. W. "Durable Secondary Reinforcement: Method and Theory." *Psychological Review* 64 (1957): 373–83.

INDEX